DEVELOPMENT IN MAMMALS
VOLUME 4

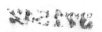

DEVELOPMENT IN MAMMALS

VOLUME 4

Editor

Martin H. Johnson

Lecturer in Anatomy, University of Cambridge
Fellow Christ's College, Cambridge

1980

ELSEVIER/NORTH-HOLLAND BIOMEDICAL PRESS
AMSTERDAM · NEW YORK · OXFORD

ISBN for the series: 0 7204 0632 3
ISBN for this volume: 0-444-80274-6

Published by:

Elsevier/North-Holland Biomedical Press
335 Jan van Galenstraat, P.O. Box 211,
Amsterdam, The Netherlands.

Sole distributors for the U.S.A. and Canada:

Elsevier North-Holland Inc.
52 Vanderbilt Avenue
New York, N.Y. 10017.

Library of Congress Cataloging in Publication Data (Revised)
Main entry under title:

Development in mammals.

    Includes bibliographies and indexes.
    1. Embryology--Mammals. I. Johnson, M. H.
QL959.D36   599.03'3   77-5894
ISBN ʺ 0-444-80274-6 (Vol. 4)

Printed in the Netherlands.

PREFACE

Volumes 4 and 5 of Development in Mammals follows the same broad framework as before. The collection of contributions pursues aspects of positional recognition in developmental systems, but neither exclusively nor comprehensively. In this volume, there are contributions on inductive systems, contributions on systems of pattern formation and morphogenesis and several, more general, contributions which discuss mechanisms that might operate in both of these types of system.

I wish to thank Shirley French and Raith Overhill for their help and the contributors for their cooperation.

30th April, 1980

Martin Johnson
Department of Anatomy
Downing Street
Cambridge

CONTENTS

© 1980 Elsevier/North-Holland Biomedical Press
Development in Mammals, Vol. 4, M.H. Johnson, editor

POSITION IN THE EMBRYO

Martin Johnson

Department of Anatomy
Downing Street,
Cambridge, U.K.

It is plausible to propose that embryogenesis involves
the unfolding within the embryo of a temporal programme of
information, components of which are selected for implementation
according to the position a cell finds itself in. The evidence
that position may play a decisive role in influencing cell fate
is old and extensive. It takes two broad forms which are
usually interpreted in terms of the theories of induction and of
pattern formation. The two distinct terminologies help to mask
what is probably a fundamentally similar mechanism at work. As
we learn more about pattern formation, so it becomes clearer
that, whatever the nature of the signals that specify positional
values, they almost certainly operate within or between small
populations of cells, which, with expansion by division, remember
and reveal their early instruction.

The generation and the recognition of positional differences
in a whole range of developmental processes thus share a common
feature - they involve short range interactions as a result of
which cells are changed either overtly or latently. Why is it
that we are still so far from understanding how this is achieved?

There seem to be three main reasons. First, instructive
events may be temporally very remote from any directly observable
consequences. This is particularly so for pattern formation.
The greater the interval between instruction and expression, the
more scope exists for distortion. Immediate measures of received
instruction are required. Second, the cell populations studied
are usually complex, heterogeneous, and probably include many
irrelevant cells which elevate biochemical noise when immediate
measures of received instruction are sought. Third, the cell

populations involved are small and therefore the techniques
required for analysis of positional differences must be made
correspondingly small. It is the solution to this third problem
which will permit an entry into the rigorous study of the other
two. How big can an instructional field be? How can we define
its operational limits? When does a single cell become big
enough to constitute a whole field? If so, can we use the
archaic, but increasingly plausible, notion of cytoplasmic zones
within a cell as a model for distinctive cellular units within
a population of cells?

These questions are not answered in this volume and I doubt
that they will be in the next which pursues this theme. The
questions are nibbled at, however. The diet is not a balanced
one, but taken together with some contributions in earlier
volumes and with the one to follow, it begins to provide some
clues as to the kind of developmental system that might yield to
experimental pressure and to the kind of technical approaches
that are needed.

© 1980 Elsevier/North-Holland Biomedical Press
Development in Mammals, Vol. 4, M.H. Johnson, editor

INTERCELLULAR COUPLING IN

MAMMALIAN OOCYTES

R. M. Moor and D.G. Cran

A.R.C. Institute of Animal Physiology,

Animal Research Station,

307 Huntingdon Road,

Cambridge CB3 0JQ,

U.K.

1.   Permeable junctions in somatic cells

     1.1 The permeable junction

     1.2 Intracellular modulation of permeable junctions

     1.3 Role of permeable junctions in cellular regulation

2.   Structure and function of cumulus-oocyte coupling

3.   Time dependent changes in cell coupling during oocyte
     maturation

4.   Hormonal control of intercellular coupling in oocytes

     4.1 Follicle size and cumulus-oocyte coupling

     4.2 Purified gonadotrophins and cumulus-oocyte coupling

     4.3 Steroids and cumulus-oocyte coupling

5.   Intracellular requirements for cumulus-oocyte coupling

     5.1 Inhibition of intracellular synthesis and
         cumulus-oocyte coupling

     5.2 Microtubules, microfilaments and oocyte-cumulus
         coupling

6.   Regulatory mechanisms and conclusions

     Many somatic cells form permeable junctions through which
small molecules pass freely from one cell interior to another.
These junctions have been associated with the direct transmission
of electrical and molecular signals that are thought to regulate
and synchronise cell function in multicellular systems (Furshpan
& Potter, 1959; Pitts, 1972; Gilula et al, 1972; Bennett, 1977;
Loewenstein, 1979; Lo, this volume, and others). An elegant

example of this form of cellular regulation has been the
demonstration that hormonal stimulation can be communicated
between granulosa and myocardial cells by contact dependent
mechanisms (Lawrence et al, 1978). A similar but as yet
unsubstantiated role has also been postulated for the permeable
junctions that exist between cumulus cells and oocytes in mammals
(Anderson & Albertini, 1976).

The purpose of this contribution is to consider the
structural basis, function and regulation of junctional communication between somatic and germ cells in the mammalian ovary.
To establish a framework for this discussion, we begin by briefly
outlining some of the most salient features about intercellular
coupling in somatic cells. Comparisons of somatic-to-somatic
and somatic-to-germ cell coupling reveal differences in the
structure of the communicating system in the two models. New
findings on the regulation of cell coupling are presented and
discussed with reference to the structural specialization of the
cumulus-oocyte junction. We conclude by speculating on the role
of permeable junctions in oocyte maturation.

1. PERMEABLE JUNCTIONS IN SOMATIC CELLS

1.1 THE PERMEABLE JUNCTION

Evidence accumulated from many different studies and
critically evaluated by Griepp and Revel (1977) and Loewenstein
(1979) suggests strongly that gap-junctions form the principal
intercommunicating system between somatic cells. These structures,
seen most clearly in freeze-fracture electron microscopy, are
arrayed in the plane of the membrane and are often found in the
form of a hexagonal lattice. A penetrating insight into the
structure of the individual gap junction is provided by the
papers of various authors (Unwin & Zampighi, 1980; Markowski
et al, 1977; Caspar et al, 1977). From an analysis of X-ray
diffraction, ultrastructural and chemical information, these
authors concluded that mouse liver gap junctions consist of a
central channel bounded by six pairs of tightly joined hexameric
proteins with a central water core. The precise alignment of the
subunits and the close apposition of the bounding proteins provide a firm structural basis for the unit channel hypothesis of

Loewenstein (1966). According to this hypothesis, the permeable junctions consist of numerous discrete and leakproof hydrophilic channels spanning the lipid bilayer. The junctional unit concept suggests further that all membrane channels in a junctional complex are similar and allow passage of all sizes of molecules within the permitted size range which, in mammals, is limited to 800 daltons or less (Flagg-Newton *et al*, 1979).

1.2 *INTRACELLULAR MODULATION OF PERMEABLE JUNCTIONS*

The means by which cells regulate junctional permeability is fraught with uncertainty. Loewenstein (1979) points out that the rate of molecular or electrical passage through permeable junctions depends upon three factors, namely, (i) the chemical or electrical gradients between the generator and receptor cell, (ii) the frictional factors or conductance within individual junctional channels, and (iii) the number of functional unit channels. A complete characterization of the extent to which each of these factors contribute to a change in signal transmission between cells is difficult to determine. This in turn reduces the accuracy with which the mechanisms responsible for specific changes to the membrane channel can be identified. However, it has been established that an increase in intracellular levels of ionized calcium reduces or totally abolishes junctional permeability in a wide variety of cells (reviewed by De Mello, 1975; Loewenstein, 1979; Flagg-Newton & Loewenstein, 1979). This has led to the formulation of a unified hypothesis which postulates that factors which alter junctional permeability all act by altering the concentration of Ca^{2+} near the junction (Loewenstein, 1966, 1967). An expansion of this model of junctional control has been advanced by Hax *et al.* (1974) and van Venrooij *et al.* (1974). The Ca hypothesis has been used to explain the means by which such varied agents as metabolic inhibitors, CO_2 and ionophore A23187 depress junctional permeability in mammalian cells (Flagg-Newton & Loewenstein, 1979). This hypothesis is, however, challenged by Turin and Warner (1977) who argue that changes in cytoplasmic pH rather than Ca^{2+} provide the final mechanism for the regulation of junctional permeability.

1.3 ROLE OF PERMEABLE JUNCTIONS IN
CELLULAR REGULATION

Numerous theories have been formulated about the role of permeable junctions in cellular regulation. It has been suggested that cell-to-cell coupling is vital in early development and differentiation, in nervous and cardiac function, in malignancy, in the modulation of hormonal function and in many other functions (reviewed by Bennett, 1977; de Mello, 1977; Griepp & Revel, 1977; Pitts & Finbow, 1977; Larsen, 1977; Loewenstein, 1979; and others). Although little would be gained in this paper by a detailed consideration of these varied theories they nevertheless serve to underline two broad conclusions about the functional significance of intercellular coupling. It is firstly clear that surprisingly little direct experimental data exist on the role of permeable junctions in cell biology. Secondly, it would appear probable that permeable junctions transmit two broad classes of molecules, namely those with an instructional and those with a permissive or nutrient role. Interpretation of results is complicated by the fact that any change in junctional permeability is likely to affect trans-mission of both of the above classes of molecules.

2.   STRUCTURE AND FUNCTION OF CUMULUS-OOCYTE COUPLING

During much of follicular development the oocyte maintains a direct structural relationship with the cells which surround it. In primordial and primary follicles the epithelium of the pre-granulosa abuts closely upon the oocyte plasma membrane and interdigitation between the membranes of the two cell types is common. In some species, e.g. rhesus monkey (Zamboni, 1974) and mouse (Anderson, 1972), desmosomes connect the epithelial cells to the oocyte. In a somewhat similar manner desmosomes also connect somatic cells to each other. To date only one report (Anderson & Albertini, 1976) has described the presence of gap junctions between follicle cells and oocytes in these immature follicles. Similar gap junctions may also occasionally connect somatic cells in immature follicles (Fletcher, 1978) but extensive proliferation of these junctions does not occur until a later developmental stage (see below).

As the follicle and oocyte start enlarging the zona pellucida is formed. This structure imposes a continuous barrier between somatic and germ cells. Specialized processes, produced by the cells adjacent to oocyte, penetrate the zona and make contact with the oocyte surface. In antral follicles, the processes are a prominent and unique feature of the cumulus-oocyte coupling system and are seen in the light microscope as transzonal striations (Figures 1 and 2). Information regarding the number of such processes is largely lacking and for species in which much branching occurs (Zamboni, 1974; Anderson & Albertini, 1976) would be difficult to obtain. In the sheep, in which little branching takes place, direct measurement of thick epon sections through the oocyte has indicated that there are in the region of 8000 processes. This in turn indicates that each of the cells immediately adjacent to the oocyte subtends some seven or eight processes.

The processes are generally devoid of organelles and are consistently filled with a compact array of microfibrils orientated in their long axis (Figure 3a). The filaments do not anchor onto the cell membrane nor do they aggregate on a specific structure within the cells from which the projections originate (Figure 4a). The processes generally terminate in a bulbous swelling directly on the oolemma and do not have any interaction with villi which may arise from it. However, in some species, a clustering of villi around the point of entry has been reported (e.g. Zamboni, 1974). The bulbous projections appear to be a specialized feature of somatic-germ cell interaction and have no obvious counterpart in somatic-somatic cell coupling. Such a swelling clearly increases the surface area of the heterologous contact. The process may terminate directly on the surface or more frequently 0.5 to 1.0μm within the ooplasma (Figures 2 and 3a). In the rhesus monkey (Hope, 1965) and human (Baca & Zamboni, 1967), the processes may penetrate particularly deeply into the oocyte cytoplasm.

In all species so far examined, a junctional complex which is morphologically of the intermediate junction (fascia adherens) type has been observed between the process terminals and the oolemma (Figure 3b). This junctional formation is characterized

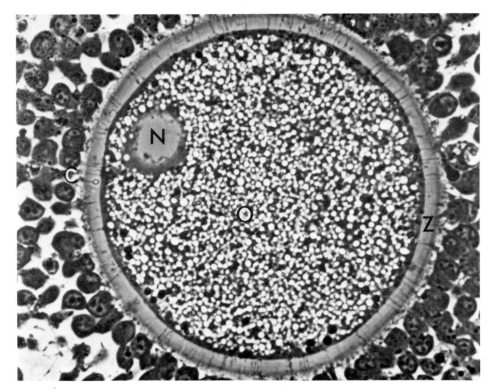

Figure 1: Section through an oocyte (O) from an antral follicle demonstrating the presence of processes from cumulus cells (C) traversing the zona pellucida (Z). The nucleus (N) is in the dictyate stage. X 760.

by electron density on the cytoplasmic side of the adjacent membranes and is similar to that observed between many somatic cell types (e.g. Rhodin, 1974). In the sheep (Cran et al, 1980) an electron dense discontinuous line is observed in the intervening space. With freeze fracture techniques, large intramembrane particles have been observed on the outer membrane leaf (E fracture face) in the desmosomal regions although a similar arrangement has yet to be identified in the inner membrane leaf or P fracture face (Gilula et al, 1978). There is a general current concensus of opinion that such desmosomal structures have a stabilizing function in anchoring together the two cell types (Szollosi et al, 1978; Gilula et al, 1978).

During recent years the observation of gap junctions at the point of heterologous cell contact (Amsterdam et al,1976;

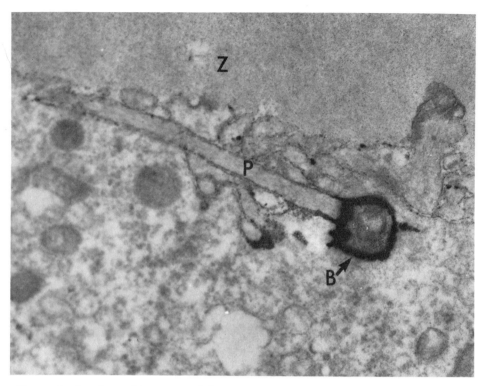

Figure 2: Cumulus cell process (P) outlined with lanthanum, entering an oocyte and showing the bulbous base (B). Z - zona pellucida. X 24,000.

Anderson & Albertini, 1976; Gilula, 1977; Gilula *et al*, 1978; Szollosi *et al*, 1978; Fletcher, 1979) has prompted much interest. This is due to their involvement as a site of structural pathway for cell-to-cell communication in a manner analogous to that between somatic cells (see previous section). Gap junctions between cumulus cells and oocytes are small, seldom exceeding 100-200 nanometers. This is in marked contrast to those between granulosa and cumulus cells where the gap junctions are large, occasionally persisting for several micrometers, as seen in thin sections. Heterologous gap junctions are formed by a direct interaction between the oocyte surface and the cumulus cell processes with a 2 - 4nm intervening 'gap' and are located at a position within the intermediate junctions. To date, the number of species in which unequivocal heterologous gap junctions have been detected is small. This is somewhat surprising in view of

10

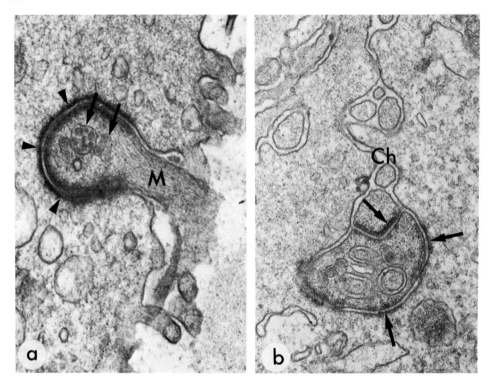

Figure 3: (a) Bulbous end of a cumulus cell process. The process is filled
with microfibrils (M) and contains a few membranous elements (arrows). An
extensive intermediate junction is present (arrowheads). X 39,000. (b) Base
of a cumulus cell process with a deeply penetrating channel (Ch) from the
oocyte surface. Several intermediate junctions are present (arrows). The
intercellular space at the junctions contains a discontinuous central layer of
material. Z - zona pellucida. X 20,000. (From: Cran *et al*, 1980).

the wide species range in which the oocytes have been examined at
an ultrastructural level and has been attributed to their small
dimensions. The authors have examined thin sections of sheep
oocytes with a view to demonstrating gap junctions. Although
extensive gap junctions were observed between granulosa and
cumulus cells, even after lanthanum impregnation it has not proved
possible to demonstrate unequivocally junctions with a similar
appearance to those between somatic cells. It is likely that
recourse to freeze fracture must be made for a definitive demonstration.

As seen with freeze fracture, the gap junctions may take
several forms (Anderson & Albertini, 1976; Gilula *et al*, 1978;

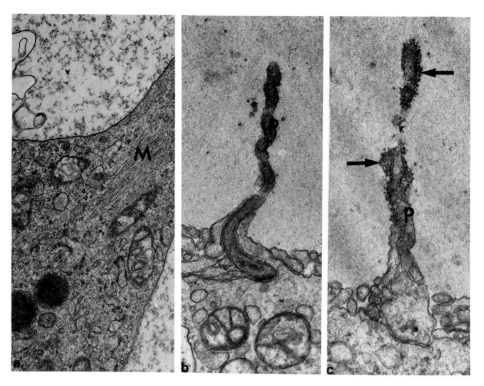

Figure 4: (a) Microfibrils (M) within a cumulus cell process at the point of
initiation. Follicle treated with FSH. X 14,000. (b) Part of a process (P) in
an untreated follicle. X 18,500. (c) Process (P) in a follicle treated with
FSH for 9 h. Note the presence of numerous small vesicules (arrows). X 20,000.

Fletcher, 1978) from only two or three 8 - 9nm particles to
macular aggregates of 100 or more particles. They may be observed
as linear or polymorphic aggregates and strands. Since gap
junctions probably represent the site of intercellular molecular
transfer, it is of interest to attempt to calculate the proportion
of the oocyte surface they occupy. From the limited number of
freeze etch micrographs in the literature, it can be calculated
that approximately 2% of the terminal process surface is occupied
by gap junctions. With reference to the previous data given for
the sheep oocyte, it seems likely that, in spite of their high
frequency, less than 1% of the total surface area of an oocyte
possesses gap junctions. Calculations of this nature are, of
course, imprecise; however, they do serve to underline the fact
that if, indeed, intercellular transfer to the oocyte occurs at

this site, the rate of such movement is likely to be very
high.

Evidence that oocytes and their surrounding cumulus cells
are physiologically coupled comes from two sources; electrotonic measurements and tracer studies.

Bidirectional ionic coupling has been demonstrated using
microelectrodes positioned in the oocyte and in an adjacent
cumulus cell (Gilula et al, 1978). This method provides a
sensitive test for the presence of junctional channels of a size
which allows passage of those inorganic ions that carry the test
electrical current through the junctions. The passage of
inorganic ions does, however, not necessarily imply that the
junctions are permeable to larger molecules (Flagg-Newton &
Loewenstein, 1979).

The use of fluorescent tracers, iontophoresed into the donor
cell, provides direct quantitative information about junctional
permeability (see Loewenstein, 1979; Lo, this volume). Dye
transfer from the oocyte to the surrounding cumulus cells has
been demonstrated in mammals (Gilula et al, 1978) and amphibia
(Browne et al, 1979).

Radioactive tracers have been used in autoradiographic and
biochemical studies to measure metabolic coupling between cells
(see Pitts & Finbow, 1977). These studies depend upon the uptake
and conversion of radiolabelled precursors into intermediate
metabolites which cannot thereafter move across the cell membrane.
We have used this procedure to quantify the extent of coupling
between the ovine oocyte and its surrounding cumulus cells.
For the autoradiographic studies granulosa cell monolayers were
labelled with either ^{3}H choline or uridine and then washed
extensively to remove the non-metabolized precursor from the
cultures. Cumulus masses with an associated oocyte were thereafter co-cultured with the labelled granulosa cells for one to
three hours. In most cases, the cumuli attached to the underlying pre-labelled cells and autoradiographic procedures showed
that the labelled intermediates had been transmitted into the
cumulus cells and oocyte (Figures 5 and 6). On the other hand,
if attachment of the cumulus did not occur or was mechanically
prevented by the interposition of a millipore filter, no transfer

of labelled products occurred. This indicates that the formation
of intercellular junctions is necessary for the passage of
labelled intermediates into the recipient cumulus cells.

Choline has been used for most of the more quantitative biochemical determinations of the effect of hormones and other agents
on junctional transmission into ovine oocytes. Choline has a low
molecular weight well within the maximal permitted size for
junctional passage (Flagg-Newton *et al*, 1979). It is not trans-
ported across the oolemma and is therefore excluded from direct
entry into oocytes from which the surrounding cumulus cells have
been removed. However, its uptake into cumulus cells is rapid
and after phosphorylation choline metabolites cannot leave the
cells except through permeable junctions. The passage of labelled
choline metabolites into cumulus enclosed oocytes provides a
measure of metabolic coupling between the two cell types.

3. TIME-DEPENDENT CHANGES IN CELL COUPLING DURING OOCYTE MATURATION

It is clear from the foregoing discussion that oocytes and
their associated follicle cells form permeable junctions which
provide for the direct flow of small molecules between somatic
and germ cell interiors. The intercommunication between these
two cell types has provoked much speculation about the role of
permeable junctions as regulators of meiotic maturation in
mammals (Anderson & Albertini, 1976; Dekel & Beers, 1978; Gilula
et al, 1978). The evidence for and against the various theories
of junctional regulation of meiosis will be evaluated in the final
section of this paper. It is important to stress at the outset,
however, that much of the present conjecture and speculation
about the role of oocyte-cumulus coupling lacks direct experimental support. Useful information can be obtained nevertheless
by considering the temporal relationships that exist in oocytes
between the intracellular changes during maturation and the
associated changes in intercellular coupling.

In rabbits the chronology of meiotic change in oocytes
matured *in vivo* or *in vitro* has been determined by Chang (1955)
and Thibault and his colleagues (1976). It was found that the
breakdown of the germinal vesicle and the resumption of meiosis
occurred about 3.5 hours after coitus or 2.5 hours after LH

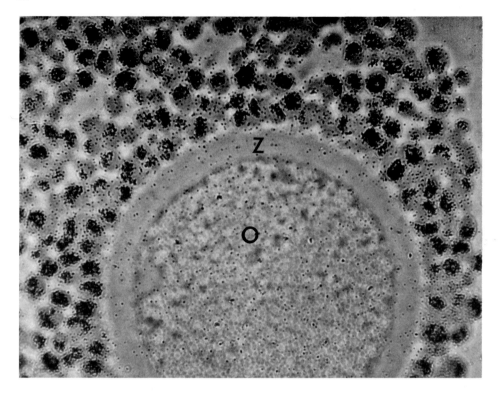

Figure 5: Autoradiograph of a cumulus mass incubated in tritiated uridine.
Silver grains are present   throughout the cumulus (C) and over the zona
pellucida (Z) and oocyte (O). X 830.

release. The initiation of metaphase 1 occurred 3.5 hours after
LH and the initiation of the second metaphase spindle occurred at
about 8.5 hours after LH. More recent experiments have provided
information on the time-related changes in protein synthesis in
rabbit oocytes during maturation (Van Blerkom & McGaughey, 1978).
No detectable changes occur in the profile of polypeptides during
the first 3 hours after the injection of HCG into oestrous does.
Between 3 and 6 hours after HCG treatment, synthesis of certain
polypeptides ceases and the synthesis of new polypeptides is
initiated. In addition, further major changes occur in the poly-
peptide profile in the 6 hours prior to ovulation. Changes in
cumulus-oocyte coupling in this species have been studied by
Szollosi and his associates (Szollosi, 1975; Szollosi *et al*, 1978)
In these investigations, ultrastructural techniques were used to

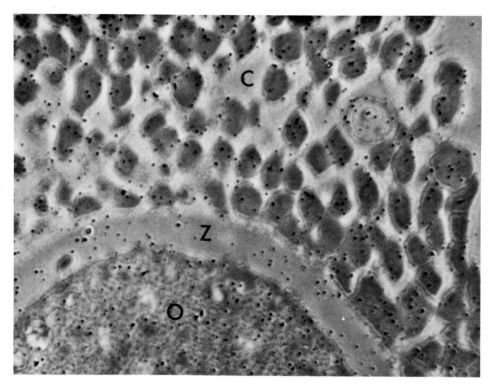

Figure 6: Autoradiograph of a cumulus mass cultured on a granulosa layer preincubated with tritiated choline. The silver grains show a similar disposition to that in Figure 5. X 970.

identify junctional changes in oocytes obtained from rabbits at 30 to 60 min intervals over the 12 hour period between mating and ovulation. No disruption to intercellular communication occurred until about 5 hours after mating when widespread uncoupling of the cumulus cells and oocytes occurred. It would therefore appear probable that changes in the nucleus and cytoplasm of the maturing rabbit oocyte occur before the communication between the cumulus and oocyte is disrupted.

In the rat oocyte, disappearance of the nucleolus and breakdown of the germinal vesicle occurs within 2 hours of the gonadotrophin surge (Tsafriri & Kraicer, 1972). The first metaphase spindle is formed at 5 - 6 hours and the second metaphase forms at 8 - 9 hours after the release of LH (Mandl, 1963). No measurements of protein synthesis in preovulatory rat oocytes have yet been published. Electrophysiological measurements have

demonstrated that ionic coupling between cumulus cells and
oocytes is fully maintained up to 2 hours after the injection of
hCG to suitably primed female rats (Gilula et al, 1978). Indeed,
disruption of ionic coupling during oocyte maturation appears
to be a relatively late event, when compared with changes in the
nucleus. For instance, at 5 hours after LH, cumulus-oocyte
coupling was undisturbed in 64% of instances and it was not until
7 hours post LH that the majority of oocytes became uncoupled.
It must, however, be recalled that changes in ionic coupling are
not necessarily indicative of similar changes in the passage of
larger molecules. The possibility therefore exists that the
passage of instructional molecules to rat oocytes changes at
times other than those outlined above.

The period from LH release to ovulation in sheep, namely
22-24 hours, is approximately twice as long as that for rodents
or rabbits. The breakdown of the germinal vesicle occurs by
9 hours after LH release, the first metaphase spindle forms at
12-15 hours and the second metaphase spindle forms about 6 hours
later (Dzuik, 1965). Corresponding changes in the profile of
polypeptides can be demonstrated. Synthesis is unchanged during
the first 6 hours after LH but by 9 hours marked changes are
apparent. These changes are characterized by the disappearance
of certain proteins seen during the early phase of maturation and
the synthesis of new or 'late stage' maturation proteins (Warnes
et al, 1977).

The intercellular passage of choline and its metabolites into
ovine oocytes remains at a consistently high level during the
first 12 hours after the initiation of maturation by LH injection
(Figure 7). The entry of marker choline into the oocyte declines
sharply in the ensuing 3 hours to reach 40% of its former rate
by 15 hours after LH; there is then no further significant
reduction in cumulus-oocyte coupling before ovulation. This
reduction in junctional transmission is matched by a significant
increase in amino acid flux across the oocyte membrane (Moor &
Smith, 1979). Biochemical, autoradiographic and ultrastructural
studies combine to show that the reduced junctional transmission
of choline 12-15 hours after LH is due to disruptive changes in
the cell processes and junctions rather than to intracellular

changes in choline metabolism. It is noteworthy that the
decline in the intercellular passage of the second marker, namely
uridine, occurs between 9 and 12 hours after the administration
of LH (Moor *et al*, 1980). The difference in the time at which
the entry of choline and uridine into the oocyte is reduced after
LH could be indicative of junctional selectivity: selective
changes in junctional permeability have been observed in somatic
cells treated with ionophore A23187 or CO_2 (Flagg-Newton &
Loewenstein, 1979). Equally, differences between the two markers
could result from changes in the rates of metabolism of uridine
in either the cumulus cells or oocyte during the maturation process. Our experiments do not enable us to choose definitively
between the two possibilities at present. Despite these differences, it is apparent that maturational changes in the nucleus

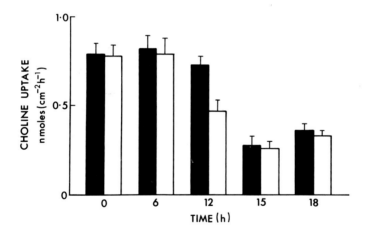

Figure 7: Extent of intercellular coupling between cumulus cells and oocytes
at selected stages after gonadotrophin treatment. Cumulus-oocyte complexes
were removed from pre-ovulatory follicles *in vivo* ■ or from large non-
atretic follicles maintained in culture □ . Transmission of labelled marker
into oocytes was determined after 60 min incubation of cumulus-oocyte masses
in 30 µM [³H] choline. Adapted from Moor *et al.* (1980).

and cytoplasm of ovine oocytes, in common with the other species
so far examined, precede measurable changes in the degree of
intercellular coupling between the cumulus and the oocyte.

It is interesting that the functional characteristics of
intercellular communication in amphibian oocytes appear to be the
reverse of those seen in mammals. Before the induction of
maturation in *Xenopus laevis*, fluorescein dye injections show
that few if any permeable junctions exist between the oocyte and
its associated follicle cells (Browne *et al*, 1979). The administration of gonadotrophin in amphibia initiates meiotic
maturation and stimulates the opening or assembly of functional
gap junctions. It is postulated that the development of this
hormonally-induced intercellular coupling system is necessary for
the transmission of specific meiotic initiator factors directly
into the oocyte cytoplasm.

4. HORMONAL CONTROL OF INTERCELLULAR COUPLING IN OOCYTES

To maintain physiological relevance, measurements of intercellular coupling during maturation have been made on oocytes
obtained directly from the animal. Identification of the underlying regulatory mechanisms have, however, necessitated the study
of cell coupling in oocytes from follicles exposed to specific
agents *in vitro*. Of critical importance in such experiments is
the selection of conditions which ensure that cell function in
culture is similar to that pertaining *in vivo*. That these conditions can be achieved *in vitro* is suggested by the following
observations. Oocytes maintained within the follicle and
cultured with appropriate hormones and under hyperbaric conditions fulfil the most important and stringent criterion of
acceptability, namely that they develop the capacity to undergo
normal fertilization and fetal development when transferred to
host females (Thibault *et al*, 1975; Moor & Trounson, 1977). In
addition, intercellular coupling in oocytes matured in large
antral follicles *in vitro* changes in a manner which approximates
closely that observed *in vivo* (see Figure 7).

The results, largely as yet unpublished, to be discussed in
the remainder of this paper have been obtained using oocytes
removed from follicles after 15 hours in culture (see Moor &

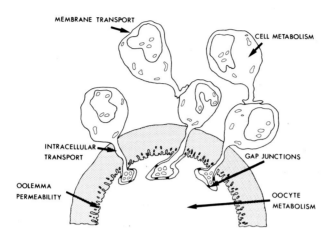

Figure 8: Diagrammatic representation of the cumulus-oocyte complex showing the intracellular sites at which hormones and other agents might disrupt inter-cellular transmission of metabolic products.

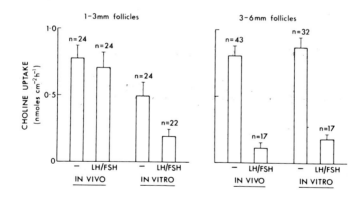

Figure 9: Intercellular transmission of choline into oocytes obtained from small (1-3mm diameter) or large (3-6mm diameter) follicles. Cumulus-oocyte complexes were removed (i) before gonadotrophin release (-) or 15 h after gonadotrophin release *in vivo* (LH/FSH group) or (ii) from follicles cultured *in vitro* for 15 h in the absence (-) or presence (LH/FSH) of LH, 3µg; FSH 5µg; prolactin 20mg ml-1 medium. Uptake was determined after 60 min incubation of cumulus-oocyte masses in 30µM [^{3}H] choline.

Trounson, 1977, for methods). The subsequent incubation of
oocyte-cumulus complexes with [^{3}H] choline has been carried out
according to the method of Moor *et al.* (1980). The results are
expressed as nmoles cm^{-2} h^{-1}. Uptake of marker is calculated
as the amount of radioactive precursor in the oocyte, irrespective of whether or not this has been further metabolised. The
uptake is expressed in terms of oocyte surface membrane area
(see Moor & Smith, 1978).

Caution must be exercised in the interpretation of intercellular flux studies because of the numerous sites, outlined in
Figure 8, at which passage of physiological substances could
possibly be modulated. For example, entry of labelled marker
into a recipient cell could be altered by modifying membrane
function, rate of synthesis, metabolism or movement of the marker
in the donor or generator cell. Equally, alterations to
junctional permeability or changes in the metabolism or function
of the receptor cell could also affect the amount of labelled
marker entering that cell. To assist in the interpretation of
the data, we have in all cases carried out studies in which both
the extent of cell-to-cell transfer of marker and the ultrastructure of the cumulus-oocyte complexes have been determined.
In addition, biochemical measurements of metabolic events taking
place in both donor and recipient cells have been carried out in
some experiments.

4.1 *FOLLICLE SIZE AND CUMULUS-OOCYTE COUPLING*

It is a consistent finding that the release of gonadotrophin
in vivo sharply reduces junctional permeability in oocytes from
large preovulatory follicles but does not depress transmission in
smaller follicles (Figure 9). Moreover, gonadotrophins added to
large follicles *in vitro* reduce choline transmission to the same
extent as that observed *in vivo* (see above, plus Figures 7 and 9).
Of note, however, is the greatly increased sensitivity of small
follicles in culture to gonadotrophin; intercellular coupling in
oocytes from small follicles *in vitro* is as disrupted by gonadotrophin as that in large follicles (Figure 9). Data on gonadotrophin receptors, steroid function and ovarian tissue interactions
are being studied in an attempt to identify the reasons for the
enhanced junctional sensitivity of small follicles *in vitro*. Only

large follicles (>3.5mm in diameter) have however been used in the work in this communication.

The mechanisms by which gonadotrophin reduces intercellular transmission of metabolites in oocytes from large follicles have been examined using biochemical and electron microscopical techniques. Metabolic studies indicate that the gonadotrophin-induced decline in coupling is not caused by a change in choline metabolism in either the cumulus cells or the oocyte (Moor *et al*, 1980). Electron microscopical studies suggest further that the decline in intercellular coupling after gonadotrophin treatment may not be due to changes in the permeable junctions themselves. It seems more probable that the primary lesion occurs within the processes of the cumulus cells that terminate in junctional contact with the oolemma (Gilula *et al*, 1978; Cran, D.G. & Moor, R.M., unpublished observations). Before gonadotrophin release only a few vesicular elements are detected in the bulbous base of the processes. However, following release, the intermediate junctions between the processes and the oolemma disappear and membrane-bound organelles with a structure similar to either lysosomes or multivesicular bodies are observed. In addition, numerous small vesicles of unknown function are found clustered around the processes as they pass through the zona pellucida (Cran *et al*, 1980).

4.2 *PURIFIED GONADOTROPHINS AND CUMULUS-OOCYTE COUPLING*

The contribution of the individual gonadotrophins to the disruption of intercellular coupling between the cumulus cells and oocyte is summarised in Figures 10, 11 and 12 (Moor *et al*, 1980). Addition of highly purified LH or HCG at concentrations of up to $3\mu g\ ml^{-1}$ of culture medium fails to reduce significantly the subsequent intercellular passage of choline into oocytes (Figure 10). The flux measurements are supported closely by ultrastructural studies which show that neither the junctions nor the trans-zona processes undergo degenerative change after LH treatment.

The effect of FSH contrasts sharply with that of LH; concentrations of FSH as low as $50ng\ ml^{-1}$ reduce intercellular coupling to the same extent as does the mixture of gonadotrophins

(Figure 11). The cumulus cell processes show the characteristic membrane vesiculation that occurs in oocytes removed from pre-ovulatory follicles 15 hours after LH (Figures 4b,c and 13a). In addition, intermediate junctions are lost and lysosomes form within 9 hours after addition of FSH and their frequency increases with incubation time. A similar increase in lysosomal number is found within the cumulus cells. However, in this situation desmosome-like junctions and gap junctions between cells of the corona radiata are maintained which is suggestive of a selective action on hetero- rather than homo-cellular coupling (Figure 13b).

The addition of relatively low levels of prolactin (500ng ml^{-1}) does not significantly reduce the intercellular passage of choline into oocytes below that observed in follicles grown in the total absence of gonadotrophin (Figure 12). Junctional activity is, however, significantly reduced when higher levels of prolactin are added to the medium. These findings differ in degree from earlier results in which it was found that entry of choline into oocytes is reduced by the addition of 500ng ml^{-1} medium. High levels of prolactin have a totally different effect on the cellular ultrastructure from that induced by the complex of gonadotrophins released before ovulation. The structure of the process and its junctional complexes with the oolemma remains unchanged (Figure 14a). However, abnormally large quantities of glycogen accumulate within the body of the cumulus cells and the nuclei are frequently markedly lobed (Figure 14b). It is likely, therefore, that prolactin is, in some manner, having a deleterious effect on the metabolism of the cumulus cells which, in turn, probably affects indirectly choline transmission into the oocytes.

4.3 STEROIDS AND CUMULUS-OOCYTE COUPLING

Steroids, and in particular oestrogen, affect markedly the degree of intercellular coupling in the granulosa cells (Merk et al, 1972). Inhibitors of steroid biosynthesis and antibodies to oestrogen have been used in studies of the role of steroids in regulating intercellular coupling between cumulus cells and oocytes (Moor et al, 1980; Cran, D.G. & Moor, R.M., unpublished observations). The addition to the medium of 1.0mM amino-glutethimide, an inhibitor of the enzymes necessary for the con-version of cholesterol to pregnenolone, inhibits by 75% production

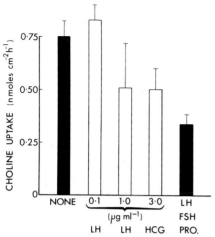

Figure 10: The effect of different concentrations of highly purified LH (LH-IRC II, courtesy Dr A. Stockell-Hartree, Cambridge) or HCG (courtesy Dr K. Bagshaw, London) on the intercellular transmission of choline between cumulus cells and oocytes. Cumulus-oocyte complexes were removed from LH-treated follicles after 15 h in culture; uptake was determined following 60 min incubation of complexes in 30μM [^{3}H] choline. LH and HCG concentrations are expressed in terms of NIH-LH-S1 equivalents.

Figure 11: The effect of different concentrations of highly purified FSH (h-FSH Seph. III, courtesy Dr. A. Stockell-Hartree, Cambridge) on the subsequent intercellular transmission of choline between cumulus cells and oocytes. Cumulus oocyte complexes were removed from FSH-treated follicles after 15 h in culture; uptake was determined following 60 min incubation of complexes in 30μM [^{3}H] choline. FSH concentrations expressed as NIH-FSH-12 equivalents.

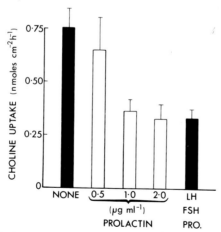

Figure 12: The effect of different concentrations of prolactin (NIH-P-S9, courtesy of N.I.A.M.D.D.) on the subsequent intercellular transmission of choline between cumulus cells and oocytes. Cumulus-oocyte comples were removed from prolactin treated follicles after 15 h in culture; uptake was determined following 60 min incubation of complexes in 30μM ^{3}H choline.

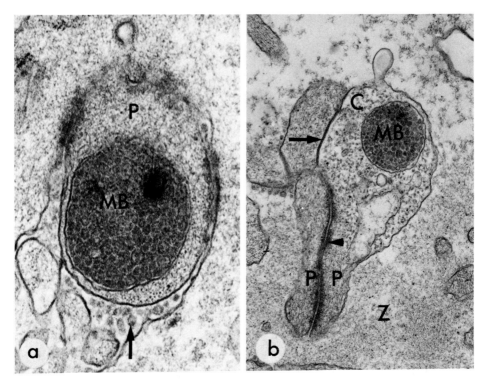

Figure 13a: A process base (P) from a follicle treated with FSH for 15 h; a
large multivesicular body (MB) is present. Small vesicles are present in the
extracellular space (arrow). X 56,500. Figure 13b: Follicle treated for
15 h with FSH. Part of a cumulus cell body (C) and of two processes (P)
entering the zona pellucida (Z). A multivesicular body (MB) is present in the
cumulus cell and a gap junction (arrow) connects the cell with the
neighbouring process. A desmosome-like junction is present between the two
processes (arrowhead). X 13,500.

of steroid by follicles in culture (Moor *et al*, 1980). The
inhibition of steroid secretion decreases slightly the extent to
which gonadotrophins suppress oocyte-cumulus coupling. On the
other hand, Figure 15 shows that the total removal of steroids
during culture using aminoglutethimide and oestrogen antiserum
depresses intercellular communication well below that induced by
FSH or other combinations of gonadotrophins. Although the role of
steroids in cumulus-oocyte coupling is still under investigation,
it seems probable that oestrogen and possibly other steroids are
necessary for the maintenance of junctional competence between
these two cell types (R.M. Moor, unpublished observations).

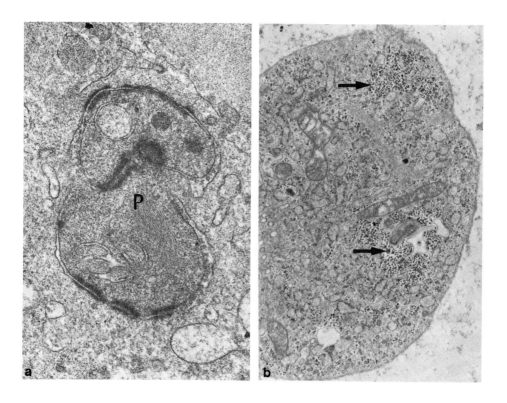

Figure 14a: A process base (P) from a follicle treated with a high level of
prolactin for 15 h. It has a structure essentially typical of that in untreated
follicles. X 30,000. Figure 14b: Cumulus cell from a follicle treated as
in Figure 14a. Large quantities of glycogen (arrows) are present. X 12,000.

5.   INTRACELLULAR REQUIREMENTS FOR CUMULUS-
OOCYTE COUPLING

The above results suggest that steroids are involved in the
maintenance of intercellular communication between cumulus cells
and oocytes in antral follicles. In addition, it is clear that
FSH is the hormone most directly involved in reducing the passage
of choline into the oocyte before ovulation. The means by which
these hormones variously support or disrupt intercellular coupling
is unclear beyond the observation that degenerative changes in
the cumulus cell processes are the first ultrastructural changes
seen after gonadotrophin treatment. Experiments carried out to
identify the sequence of intervening events between FSH binding

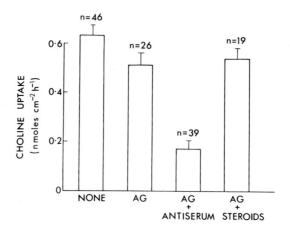

Figure 15: Effect of inhibiting follicular steroid secretion using amino-glutethimide (10^{-3}M) or antiserum to oestrogen during 15 h culture on the subsequent transmission of choline between cumulus cells and oocytes. Choline uptake was determined after 60 min incubation of cumulus-oocyte complexes in 30μM [^{3}H] choline.

and degeneration of the cell processes have as yet been uninformative. More rewarding have been the converse experiments designed to identify those intracellular mechanisms required for the maintenance of full cumulus-oocyte coupling. We have measured the extent of intercellular coupling in oocytes taken from follicles in which specific synthetic or structural com-ponents of the cells were disrupted during culture. The results are presented with the caveat that they should be interpreted with reference to Figure 8 and using the caution required in evaluating any experiments in which reputedly specific inhibitors are used.

5.1   INHIBITION OF INTRACELLULAR SYNTHESIS AND CUMULUS-OOCYTE COUPLING

The entry of choline into the oocyte is sharply suppressed by inhibiting glycosylation, RNA or protein synthesis in the follicle during culture (Table 1). The specificity of the treatments was examined by measuring certain other cellular functions in addition to junctional permeability. Thus, oestrogen secretion in follicles exposed to cycloheximide is reduced

TABLE 1:   INTERCELLULAR PASSAGE OF CHOLINE IN CUMULUS-
OOCYTE COMPLEXES OBTAINED FROM FOLLICLES
THAT HAD BEEN CULTURED FOR THE PRECEDING 15 h
WITH COMPOUNDS WHICH INHIBIT GLYCOSYLATION,
RNA OR PROTEIN SYNTHESIS

Inhibitor	Concentration $(\mu g\ ml^{-1})$	No. of oocytes	Mean ($\pm$ SEM) choline uptake by oocytes (nmoles $cm^{-2}\ h^{-1}$)
Control group	–	46	0.636 $\pm$ 0.073
α-Amanitin	10	15	0.119 $\pm$ 0.037
Cycloheximide	10	20	0.173 $\pm$ 0.042
Tunicamycin	1.0	21	0.306 $\pm$ 0.062

relative to the control group, declining from 170 $\pm$ 14 to
69 $\pm$ 7 pmoles per mg follicular tissue/15 hours in culture. On
the other hand, oestrogen secretion is increased from 170 $\pm$ 14
to 245 $\pm$ 13 pmoles mg^{-1} 15 h^{-1} by tunicamycin treatment. It is
moreover noteworthy that tunicamycin does not affect amino acid
transport or total protein synthesis in treated oocytes. Ultrastructural studies carried out in parallel with the biochemical
measurements show that none of the three treatments has any
marked affect on the structure of the processes or on the cumulus
cells. This finding is indicative that depression of intercellular communication in these cases is due to different
mechanisms from those pertaining after gonadotrophin treatment.

It should be recalled that the effect of inhibiting protein
synthesis on junctional permeability in somatic cells differs in
different cell types. Thus, protein synthesis is not required
for the maintenance of ionic coupling in Novikoff hepatoma cells
(Epstein *et al*, 1977). By contrast, assembly of junctions in
ependymal cells *in vivo* is directly dependent upon continuous
protein synthesis and upon RNA synthesis for the first 12 hours
of the assembly process (Decker, 1976).

5.2   *MICROTUBULES, MICROFILAMENTS AND OOCYTE-CUMULUS COUPLING*

Microtubules and microfilaments are important components of
an intracellular system for the rapid and directed transport of

particles and organelles in many somatic cells. Inhibitor agents have been used in our experiments as a first step in determining whether either of these cytoskeletal elements serve as a directional or force-generating system in the intercellular transfer of substances into the oocyte. That the microtubule-filament system may be of particular importance in cumulus-oocyte coupling is suggested both by the structure of the cumulus cell processes and by the requirement for closely directed movement of elements down the processes to the small junctional complexes.

Microfilaments are the most abundant structures in the cell processes traversing the zona (see Figure 4a). It has been repeatedly shown that these filaments, consisting of actin and associated proteins, form the contractile component of the cortex of somatic cells and are extremely labile (see Bray, 1973; Pollard, 1975; Lazarides, 1976, and others). Their assembly can, moreover, be effectively blocked by cytochalasin which acts on a small nucleating fragment of actin required for the assembly of the actin subunits into microfilaments (see Bray, 1979). It has been of interest in our work to examine the intercellular passage of compounds into oocytes exposed to the disruptive effects of cytochalasin. The D form of this drug has been used because, unlike cytochalasin, B, it does not block glucose uptake into cells (Miranda *et al*, 1974).

Our particular interest in microtubules and intercellular coupling arises from the hypothesis that intracellular particle transport is microtubule-directed (Freed & Lebowitz, 1970; Rebhun, 1972). Microtubules are helical polymeric structures composed of tubulin dimers which are in equilibrium with a pool of subunits (Olmsted & Borisy, 1973; Snyder & McIntosh, 1976). Depolymerization or disassembly of microtubules is induced most readily by tubulin-binding spindle alkaloids, the best known of which are colcemid, colchicine, vinblastine and vincristine (Wilson & Bryan, 1974; Wilson *et al*, 1974). The assembly of microtubules is not affected by any of the cytochalasins except cytochalasin A (Weihing, 1978). In using cytochalasin D to block microfilament formation and colcemid to depolymerise microtubules little significant cross-reaction would be expected.

A possible role for both microtubules and microfilaments in the intercellular transmission of choline from the cumulus cells to the oocyte is suggested from our experiments. It has been found that the entry of choline into oocytes is greatly reduced but not abolished in instances in which the follicles were cultured for the preceding 15 hours in medium containing cytochalasin D or colcemid (Table 2). These findings are similar to those of Stoker (1975) who first reported that cytochalasin B abolishes metabolic cooperation between somatic cells in culture.

TABLE 2: INTERCELLULAR PASSAGE OF CHOLINE IN CUMULUS-
OOCYTE COMPLEXES OBTAINED FROM FOLLICLES THAT
HAD BEEN CULTURED FOR THE PRECEDING 15 h IN THE
PRESENCE OF COMPOUNDS WHICH INHIBIT THE FORMATION
OF MICROFILAMENTS OR MICROTUBULES

Inhibitor	Concentration ($\mu g \ ml^{-1}$)	No. of oocytes	Mean ($\pm$ SEM) choline uptake by oocytes (nmoles $cm^{-2} \ h^{-1}$)
Control group	–	46	0.636 ± 0.073
Cytochalasin D	1.0	37	0.324 ± 0.052
Colcemid	1.0	24	0.104 ± 0.017

The means by which metabolic cooperation is blocked by cytochalasin is unclear since Loewenstein (1979) suggests that junctional permeability is unaffected by the drug. Indeed, he finds that even after 72 hours treatment with cytochalasin B, cells remain electrically coupled and retain their capacity to transmit fluorescent dyes via the junctional communicating system (Loewenstein, 1979). Our ultrastructural studies are of interest in this regard since they suggest that some cells are more affected than others by cytochalasin D treatment. The most marked structural changes in cytochalasin-treated follicles occur in the cumulus cells surrounding the oocyte. After treatment, these cells contain large disorganised masses of microfilaments which occupy an intracellular position adjacent to the zona pellucida (Figure 16). No unequivocal sign of microfilament disorganization has been detected in the cumulus cell processes. However, this may be due to the close packed configuration of filaments in the

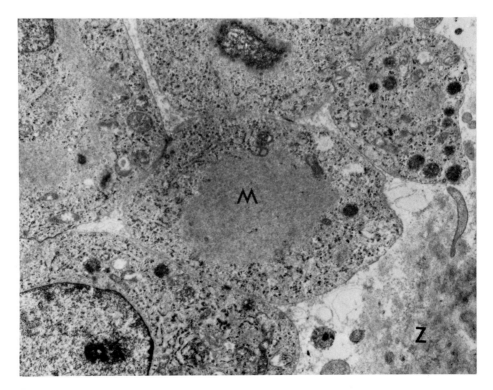

Figure 16: Follicle treated with cytochalasin D for 15 h. The cumulus cells abutting onto the zona pellucida (Z) contain masses of microfilaments (M). X 8,00.

processes. Similarly, no marked changes other than the arrest of spindle formation have been noted in the follicle cells after colcemid treatment. It is worth noting in this regard, however, that few microtubules are observed in cumulus cells even under control conditions.

6. REGULATORY MECHANISMS AND CONCLUSIONS

From the foregoing discussion it is clear that the oocyte and adjacent cumulus cells are coupled and that these cells are in turn coupled with the rest of the cumulus and granulosa. It has not been possible to obtain direct evidence that gap junctions are the site of transport. However, for somatic cells the circumstantial evidence is good and this has been assumed to apply equally to the very small junctions between the cumulus cell

processes and the oocyte. However, until this can be proven,
the whole question should be approached with an open mind,
especially since it is conceivable that selective changes in the
membrane permeability could occur at the process termini. The
precise mode of transmembrane passage of molecules is a key factor
when considering the regulation of intracellular coupling.

It has been proposed that changes in intercellular coupling.
regulate junctional permeability in a wide variety of cells
(see Flagg-Newton & Loewenstein, 1979 for references). These
workers have used this hypothesis to explain their discovery that
mammalian cells in monolayer culture are more resistant to
uncoupling than insect cells in cultures or mammalian cells in
organized tissues. Since it is unlikely that any basic dis-
similarity exists between gap junctions within different cell
types, Flagg-Newton and Loewenstein (1979) examined their monolayer cultures for other distinctive morphological attributes.
Particular attention was focussed upon the way in which cells form
contacts via gap junctions situated at the end of thin protrusions
which are devoid of mitochondria and thus lacking Ca^{2+} release
sites. It is suggested that the spatial dislocation between the
gap junction and the sites of Ca^{2+} release may account for the
resistance to junctional uncoupling in these cells.

This cellular arrangement contrasts with that between cumulus
and granulosa cells in the follicle where large areas of abutment
gap junctions are present which would be readily accessible to
high levels of intracellular calcium. However, the morphology of
cultured cells is similar to that of the cumulus cells adjacent
to the oocyte which bear long processes, generally devoid of
mitochondria, and which are connected to the oolemma by small gap
junctions. If the above hypothesis is correct, cumulus-oocyte
junctions would be more resistant to Ca^{2+} induced uncoupling than
those between cumulus and granulosa cells. However, we have
found that FSH induces cumulus-oocyte uncoupling without affecting
permeable junctions between other follicle cells. The regulation
of intercellular communication between somatic and germ cells may
therefore be different from that between adjacent somatic cells.
The cumulus cell processes have been identified as one possible
site of specialised control between oocytes and cumulus cells

(Gilula *et al*, 1978; Moor *et al*, 1980; Cran, D.G. & Moor, R.M., unpublished observations).

Intercellular communication between cumulus cells and the oocyte probably fulfils different roles at different stages of follicular and oocyte development. It may be a means whereby essential metabolic and trophic substances which are incapable of crossing the oolemma are transferred into the oocyte. Substances such as choline, uridine and inositol are of this nature. In addition, cumulus cells are important in providing necessary energy supplies for oocyte development (Biggers *et al*, 1967). This latter function may be of particular importance during the phase of oocyte growth, since it has been established that growth will only take place in contact with cumulus cells (Eppig, 1977).

Another postulated role for cell coupling is in the regulation of the nuclear and cytoplasmic changes that occur in the oocyte during maturation. In this instance the communicating system may be regarded as acting in an instructional manner rather than in the more permissive form described above.

A favoured current hypothesis suggests that an inhibitor, transmitted into the mammalian oocyte via permeable junctions, blocks meiosis at the dictyate stage (see Dekel & Beers, 1978). The disruption of the transzonal processes before ovulation would in turn lead to a decline in the intracellular levels of the inhibitor, thought to be cyclic AMP, and a consequent resumption of meiosis. The choice of cyclic AMP as the instructional molecule is based primarily on its role in the control of maturation in amphibian oocytes (Maller & Krebs, 1977). Additional evidence favouring the involvement of cyclic AMP in nuclear maturation is its capacity to regulate mitosis (Abel & Monahan, 1973), pass freely between somatic cells (Lawrence *et al*, 1978) and inhibit the occurrence of meiosis in extrafollicular oocytes in culture (Cho *et al*, 1974; Dekel & Beers, 1978). The problem with the hypothesis is that the disruption of junctional communication and the initiation of meiosis do not appear to be sequential events (see section 3 of this paper for details). It is, of course, possible that the apparent absence of a temporal relationship between the two events occurs because the current methods of measuring junctional permeability are not appropriate for measuring transmission of cyclic AMP. It would clearly be inappropriate to reject the junctional hypothesis of meiotic regulation until this technical problem is resolved.

An alternative hypothesis suggests that the disruption of cumulus-oocyte coupling initiates certain essential cytoplasmic changes in the maturing oocyte (Szollosi, 1975; Szollosi *et al*, 1978). The evidence, based on close temporal relationships, suggests that cortical granules in rabbit, cow and sheep oocytes do not become aligned directly beneath the plasma membrane until after the disruption of the permeable junctions (Szollosi *et al*, 1978; Cran *et al*, 1980). It is therefore possible that junctional control of oocyte maturation is expressed most clearly in the regulation of cytoplasmic and not nuclear events. Conclusive proof of these hypotheses will, however, depend upon more rigorous measurement of quantitative changes in the junctional communicating system in oocytes undergoing maturation.

REFERENCES

ABELL, C.W. & MONAHAN, T.M. (1973) The role of adenosine 3', 5'-cyclic monophosphate in the regulation of mammalian cell division. J. Cell Biol. 59, 549-558.

AMSTERDAM, A.R., JOSEPHS, R., LIEBERMAN, M.E. & LINDNER, H.R. (1976) Organization of intramembrane particles in freeze cleaved gap junctions of rat Graafian follicles: optical diffraction analysis. J. Cell Sci. 21, 93-105.

ANDERSON, E. (1972) The localization of acid phosphatase and the uptake of horseradish peroxidase in the oocyte and follicle cells of mammals. In: J.D. Biggers & A.W. Schuetz (Eds.), Oogenesis. University Park Press, Baltimore, Butterworths, London, pp. 87-117.

ANDERSON, E. & ALBERTINI, D.F. (1976) Gap junctions between the oocyte and companion follicle cells in the mammalian ovary. J. Cell Biol. 71, 680-686.

BACA, M. & ZAMBONI, L. (1967) The fine structure of human follicular oocytes. J. Ultrastr. Res. 19, 354-381.

BENNETT, M.V.L. (1973) Function of electrotonic junctions in embryonic and adult tissue. Fed. Proc. 32, 65-75.

BENNETT, M.V.L. (1977) Electrical transmission: a functional analysis and comparison to chemical transmission. In: E.R. Kandel (Ed.), Handbook of Physiology 1. The Nervous System. Section 1. Cellular Biology of Neurones. The Williams and Williams Company, Baltimore, pp. 357-416.

BIGGERS, J.D., WHITTINGHAM, D.G. & DONAHUE, R.P. (1967) The pattern of energy metabolism in the mouse oocyte and zygote. Proc. Natl. Acad. Sci. 58, 560-567.

BRAY, D. (1973) Cytoplasmic actin: a comparative study. Cold Spring Harbour Symp. Quant. Biol. 37, 567-572.

34

BRAY, D. (1979) Cytochalasin action. Nature, 282, 671.

BROWNE, C.L., WILEY, H.S. & DUMONT, J.N. (1978) Oocyte-follicle cell gap junctions in Xenopus laevis and the effects of gonadotrophin on their permeability. Science 203, 182-183.

CASPAR, D.L.D., GOODENOUGH, D.A., MAKOWSKI, L. & PHILLIPS, W.C. (1977) Gap junction structures. 1. Correlated electron microscopy and x-ray diffraction. J. Cell Biol. 74, 605-628.

CHANG, M.C. (1955) The maturation of rabbit oocytes in culture and their maturation, activation, fertilization and subsequent development in the fallopian tubes. J. Exp. Zool. 128, 379-405.

CHO, W.K., STERN, S. & BIGGERS, J.D. (1974) Inhibitory effect of dibutyryl cAMP on mouse oocyte maturation in vitro. J. exp. Zool. 187, 383-386.

CRAN, D.G., MOOR, R.M. & HAY, M.F. (1980) Fine structure of the sheep oocyte during antral follicle development. J. Reprod. Fert. (in press).

DEKEL, N. & BEERS, W.H. (1978) Rat oocyte maturation in vitro: relief of cyclic AMP inhibition by gonadotrophins. Proc. Natl. Acad. Sci. 75, 4369-4373.

DE MELLO, W.C. (1977) Intercellular communication in heart muscle. In: W.C. de Mello (Ed.), Intercellular Communication. Plenum Press, New York, pp. 87-126.

DZIUK, P.J. (1965) Timing of maturation and fertilization of the sheep egg. Anat. Rec. 153, 211-224.

EPPIG, J.J. (1977) Mouse oocyte development in vitro with various culture systems. Dev. biol. 60, 371-388.

EPSTEIN, M.L., SHERIDAN, J.D. & JOHNSON, R.G. (1977) Formation of low resistance junctions in vitro in the absence of protein synthesis and ATP production. Exp. Cell Res. 104, 25-30.

FLAGG-NEWTON, J. & LOEWENSTEIN, W.R. (1979) Experimental depression of junctional membrane permeability in mammalian cell culture. A study with tracer molecules in the 300 to 800 dalton range. J. Membrane Biol. 50, 65-100.

FLAGG-NEWTON, J., SIMPSON, I. & LOEWENSTEIN, W.R. (1979) Permeability of the cell-to-cell membrane channels in mammalian cell junctions. Science 205, 404-407.

FLETCHER, W.H. (1979) Intercellular junctions in ovarian follicles: A possible functional role in follicular development. In: A.R. Midgley & W.A. Sadler (Eds.), Ovarian Follicular Development and Function. Raven Press, New York, pp. 113-120.

FREED, J. & LEBOWITZ, M. (1970) The association of a class of saltatory movements with microtubules in cultured cells. J. Cell Biol. 45, 334-354.

FURSHPAN, E.J. & POTTER, D.D. (1959) Transmission at the giant motor synapses of crayfish. J. Physiol. 145, 289-325.

GILULA, N.B. (1977) Gap junctions and cell communication. In: B.R. Brinkley 8 K.R Porter (Eds.), International Cell Biology. The Rockefeller University Press, New York, pp. 61-70.

GILULA, N.B., EPSTEIN, M.L. & BEERS, W.H. (1978) Cell-to-cell communication and ovulation. A study of the cumulus-oocyte complex. J. Cell Biol. 78, 58-75.

GILULA, N.B., REEVES, O.R. & STEINBACH, A. (1972) Metabolic coupling, ionic coupling and cell contacts. Nature 235, 262-265.

GRIEPP, E.B. & REVEL, J-P. (1977) Gap junctions in development. In: W.C. De Mello (Ed.). Intercellular communication. Plenum Press, New York, pp. 87-126.

HAX, W.M.A., VAN VENROOIJ, G.E.P.M. & VOSSENBERG, J.B.J. (1974) Cell communication: a cyclic AMP mediated phenomenon. J. Membrane Biol. 19, 253-266.

HOPE, J. (1965) The structure of the developing follicle of the rhesus monkey. J. Ultrastr. Res. 12, 592-610.

LARSEN, W.J. (1977) Gap junctions and hormone action. In: B.L. Gupta, R.B. Moreton, J.L. Aschman & B.L. Wall (Eds.), Transport of Ions and Water in Animals. Academic Press, London, pp. 333-361.

LAWRENCE, T.S., BEERS, W.H. & GILULA, N.B. (1978) Transmission of hormonal stimulation by cell-to-cell communication. Nature 272, 501-506.

LAZARIDES, E. (1976) Actin, -actinin and tropomyosin interactions in the structural organization of actin filaments in nonmuscle cells. J. Cell Biol. 68, 202-219.

LOEWENSTEIN, W.R. (1966) Permeability of membrane junctions. Ann. N.Y. Acad. Sci. 137, 441-472.

LOEWENSTEIN, W.R. (1967) On the genesis of cellular communication. Dev. Biol. 15, 503-520.

LOEWENSTEIN, W.R. (1979) Junctional intercellular communication and the control of growth. Bioch. Bioph. Acta 560, 1-65.

LOEWENSTEIN, W.R., KANNO, Y. & SOCOLAR, S.J. (1978) Quantum jumps of conductance during formation of membrane channels at cell-cell junctions. Nature 274, 133-136.

MAKOWSKI, L., CASPAR, D.L.D., PHILLIPS, W.C. & GOODENOUGH, D.A. (1977) Gap junction structures. II. Analysis of the X-ray diffraction data. J. Cell Biol. 74, 629-645.

MALLER, J.L. & KREBS, E.G. (1977) Progesterone-stimulated meiotic cell division in xenopus oocytes. J. Biol. Chem. 252, 1712-1718.

MANDL, A.M. (1963) Pre-ovulatory changes in the oocyte of the adult rat. Proc. Roy. Soc. (Biol) 158, 105-118.

MERK, F.B., BOTTICELLI, C.R. & ALLBRIGHT, J.T. (1972) An intercellular response to estrogen by granulosa cells in the rat ovary; an electron microscope study. Endocrinology 90, 992-1007.

MIRANDA, A., GODMAN, G., DEITCH, A. & TANENBAUM, S.W. (1974) Actions of cytochalasin D on cells of established lines. 1. Early events. J. Cell Biol. 61, 481-500.

MOOR, R.M. & SMITH, M.W. (1978) Amino acid transport in mammalian oocytes. Exp. Cell Res. 119, 333-341.

MOOR, R.M., SMITH, M.W. & DAWSON, R.M.C. (1980) Measurement of intercellular coupling between oocytes and cumulus cells using intracellular markers. Exp. Cell Res. (in press).

MOOR, R.M. & TROUNSON, A.O. (1977) Hormonal and follicular factors affecting maturation of sheep oocytes in vitro and their subsequent developmental capacity. J. Reprod. Fert. 49, 101-109.

36

OLMSTED, J. & NORISY, G. (1973) Microtubules. Ann. Rev. Biochcm. 42, 507-540.

PITTS, J.D. & FINBOW, M.E. (1977) Junctional permeability and its consequences. In: W.C. De Mello (Ed.), Intercellular Communication. Plenum Press, New York and London, pp. 61-86.

POLLARD, T. (1975) Functional implications of the biochemical and structural properties of cytoplasmic structural proteins. In: S. Inoue & R. Stephens (Eds.), Molecules and Cell Movement. Raven Press, New York, pp. 259-286.

REHBUN, L. (1972) Polarized intracellular particle transport: saltatory movements and cytoplasmic streaming. In: G. Bourne & J. Danielli (Eds.), International Review of Cytology, Vol.32 , Academic Press, New York, pp. 93-137.

RHODIN, J.A.G. (1974) Histology, a text and atlas. Oxford University Press, New York, London, Toronto.

SIMPSON, I., ROSE, B. & LOEWENSTEIN, W.R. (1977) Size limit of molecules permeating the junctional membrane channels. Science 195, 294-296.

SNYDER, J. & McINTOSH, R. (1976) Biochemistry and physiology of microtubules. Annu. Rev. Biochem. 45, 699-720.

STOKER, M.G.P. (1975) The effects of topoinhibition and cytochalasin B on metabolic cooperation. Cell 6, 253-257.

SZOLLOSI, D. (1975) Ultrastructural aspects of oocyte maturation and fertilization in mammals. In: C. Thibault (Ed.), La Fecondation. Masson et Cie, Paris, pp. 14-35.

SZOLLOSI, D., GERARD, M., MENEZO, Y. & THIBAULT, C. (1978) Permeability of ovarian follicle; corona cell-oocyte relationship in mammals. Ann. Biol. anim. Bioch. Biophys. 18, 511-521.

THIBAULT, C., GERARD, M. & MENEZO, Y. (1976) Nuclear and cytoplasmic aspects of mammalian oocyte maturation in vitro in relation to follicle size and fertilization. In: P.O. Hubinot (Ed.), Progress in Reproductive Biology, Vol.1, S. Karger, Basle, pp. 233-240.

TSAFRIRI, A. & KRAICER, P.F. (1972) The time sequence of ovum maturation in the rat. J. Reprod. Fert. 29, 387-393.

VAN BLERKOM, J. & McGAUGHEY, R.W. (1978) Molecular differentiation of the rabbit oocyte. 1. During oocyte maturation in vivo and in vitro. Dev. Biol. 63, 139-150.

VAN VENROOIJ, G.E.P.M., HAX, W.M.A., VAN DANTZIG, G.P., PRIJS, U. & DENIER VAN DEN GON, J.J. (1974) Model approaches for the evaluation of electrical cell coupling in the salivary gland of the larva of Drosophila Lydei. The influence of lysolecithin on electrical coupling. J. Membrane Biol. 19, 229-252.

WARNES, G.M., MOOR, R.M. & JOHNSON, M.H. (1977) Changes in protein synthesis during maturation of sheep oocytes in vivo and in vitro. J. Reprod. Fert. 49, 331-335.

WEIHING, R.R. (1978) In vitro interactions of cytochalasins with contractile proteins. In: S.W. Tanenbaum (Ed.), Cytochalasins, biochemical and cell biological aspects. Frontiers of Biology, Vol.46, North Holland, Amsterdam, New York, Oxford, pp. 431-444.

WILSON, L. & BRYAN, J. (1974) Biochemical and pharmacological properties of microtubules. In: E. Du Praw (Ed.), Advances in Cell and Molecular Biology, Vol.3, Academic Press, New York, pp. 22-72.

WILSON, L., BAMBURG, J., MIZEL, S., GRISHAM, L. & CRESWELL, K. (1974) Interaction of drugs with microtubule proteins. *Fed. Proc.* 33, 158-166.

ZAMBONI, L. (1974) Fine morphology of the follicle wall and follicle cell-oocyte association. *Biol. Reprod.* 10, 125-149.

© 1980 Elsevier/North-Holland Biomedical Press
Development in Mammals, Vol. 4, M.H. Johnson, editor

GAP JUNCTIONS AND DEVELOPMENT

Cecilia W. Lo

Department of Biological Chemistry
Harvard Medical School
Boston, MA 02115

Direct cell-cell contact has been postulated to be a requirement in the differentiation of many embryonic systems. For example, in inductive tissue interactions, the differentiation of various types of epithelium (or mesenchyme) requires the close proximity of homologous or heterologous mesenchyme (or epithelium) viz. bronchial epithelium and pulmonary mesenchyme (Bluemink *et al*, 1975), enamel epithelium and odontoblastic mesenchyme (Slavkin & Bringas, 1976), and ureter epithelium and metanephric mesenchyme (Lehtonen, 1975). In particular, transfilter experiments have indicated that the induction of kidney tubule formation in the

metanephric mesenchyme probably requires direct cell-cell contact
with an inductive tissue (Lehtonen, 1976; Saxen *et al*, 1976;
Wartiovarra *et al*, 1974). Whether such interactions involve cell
surface molecules or occur via the formation of specialized cell
junctions is not known.

In this contribution I will attempt to discuss and explore
some data and ideas regarding the possible involvement of cell
junctions in the regulation of differentiation and development,
with the main emphasis being placed on the possible role of gap
junctions in early embryogenesis.

1. CELL JUNCTIONS

Three major types of cell junctions are identifiable by
electron microscopy: tight junctions (occludens junctions),
desmosomes (adherens junctions) and gap junctions (for reviews,
see Gilula, 1974; and Staehelin, 1974).

Tight junctions are characterized in thin section electron
microscopy by the apparent fusion of the outer leaflets of the
plasma membrane between two contacting cells. In freeze fracture
replicas, such areas of membrane fusion appear as 10nm ridges
on the P fracture face with their complementary grooves on the E
fracture face (Figure 1). Tight junctions provide a permeability
seal which prevents the paracellular diffusion of water and other
molecules between cells.

Desmosomes are characterized by focal areas of dense deposits
on the cytoplasmic face of the plasma membrane and a central dense
plaque in the extracellular space (Figure 2). Desmosomes are
probably involved in maintaining the mechanical stability or
anchorage between the cells of a tissue. They are part of the
"glue" which provides the necessary adhesion for maintaining
specific tissue structure and organization.

Gap junctions are typically characterized by an area of close
apposition between the plasma membrane of two cells such that the
intercellular space is reduced to only approximately 2-4nm
(Figure 3). In freeze fracture replicas, gap junctions appear as
aggregates of 8nm particles on the P fracture face with their
complementary pits on the E fracture face (Figure 4). This lattice
network sometimes can also be visualized in thin sections of
samples stained with lanthanum hydroxide (Figure 3).

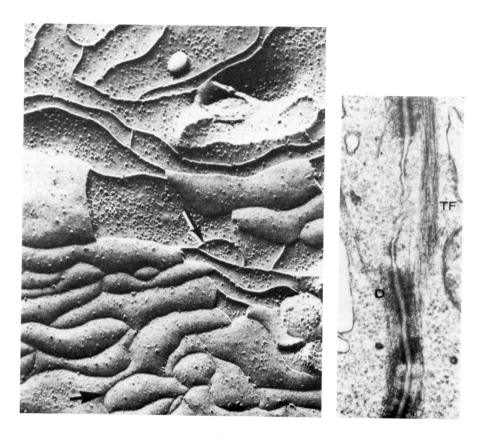

Figure 1 (above left): Freeze fracture replica revealing the tight junctional network of the trophoblast cells of the mouse embryo shortly after implantation in vitro. Note the multi-layered tiers of interconnecting ridges (black arrow) on the P fracture face and grooves on the E fracture face (black and white arrow). X 68,000.

Figure 2 (above right): Several desmosomes (D) join two adjacent differentiated endoderm-like teratocarcinoma cells. At these contacts, cytoplasmic densities along both membranes are intimately associated with 10nm tonofilaments (TF). The 30nm intercellular space contains a crystalline matrix that is bissected by a central dense lamina. X 36,000.

2. GAP JUNCTIONS AND CELL-CELL COMMUNICATION

Gap junctions are probably the structures which contain the membrane channels that mediate the direct cytoplasmic exchange of molecules between cells (Gilula *et al*, 1972; Johnson & Sheridan, 1971; McNutt & Weinstein, 1970; Payton *et al*, 1969; for reviews,

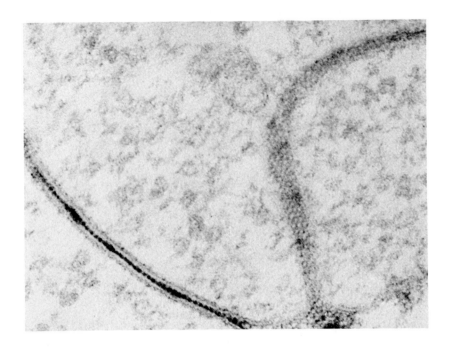

Figure 3: Gap junctions between two granulosa cells of the rat ovarian
follicle. The tissue was processed with lanthanum hydroxide. The gap junction
to the left demonstrates the typical septalaminar arrangement of plasma
membranes which are separated by a 2-4nm intercellular space. The gap junction
to the right has been sectioned en face and the penetration by the lanthanum
reveals the lattice-like arrangement of the gap junctional particles. This
arrangement is also visible from the gap junction on the left in which there
is a ladder-like pattern of alternating electron opaque and luscent areas.
X 180,000.

see Gilula, 1976; and Sheridan, 1976). The ability of cells to
communicate or to directly exchange cytoplasmic components was
first discovered in excitable neuronal cells (Furshpan & Potter,
1959) in which ionic or electrical coupling was observed. Since
that initial discovery, many other cell types have been found to
be ionically coupled (Furshpan & Potter, 1968; Loewenstein, 1966).
Such direct intercellular communication was further demonstrated
by the observation of the contact-dependent transfer of radiolabeled nucleotides during metabolic cooperation (Subak-Sharpe
et al, 1966) and the passage of microinjected fluorescent
molecules, such as fluorescein (molecular weight 330) or Porcion

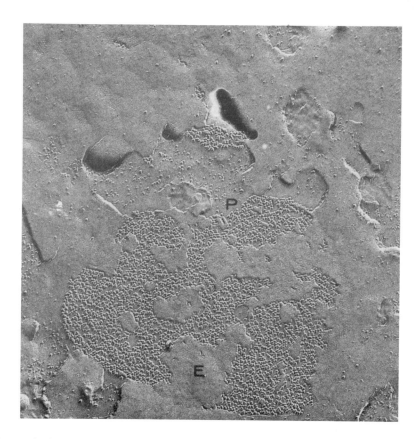

Figure 4: Freeze fracture replica of a gap junction of embryonal carcinoma cells. A plaque of particle aggregates separated by particle free aisles is present on the P face (P), while complementary pits are present on the E face (E). X 60,000.

yellow (M. W. ca. 650) (Furshpan & Potter, 1968; Johnson & Sheridan, 1971; Kanno & Loewenstein, 1966; Payton et al, 1969). Thus direct passage of molecules between cells via gap junctional communication has been detected using three basic methods: the exchange of ionic currents as probed with microelectrodes (ionic coupling) (Figure 5), the passage between cells of microinjected fluorescent dyes or other visible tracers (dye coupling) (Figure 6), and the passage of radiolabelled metabolites as visualized by autoradiography (metabolic cooperation or coupling) (Figure 7). Using a combination of these three approaches, it has been determined that gap junctions resemble passive molecular

44

sieves with a finite pore size described by an exclusion limit of
approximately 1000 daltons (Bennett, 1978; Flagg-Newton, 1979;
Pitts & Simms, 1977; Simpson et al, 1977). This conclusion
confirms a previous demonstration (Reese et al, 1971) on crayfish
segmental axons using peroxidase, a protein whose presence can be
detected after cell injection and fixation by virtue of its
enzymatic activity. It was found that coinjection of peroxidase
and microperoxidase with fluorescein resulted in the transfer of
fluorescein (M. W. 330) and microperoxidase (M. W. 1,800) into

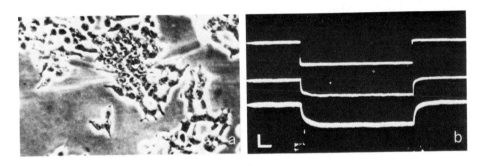

Figure 5: Ionic coupling between embryonal carcinoma cells. (a) Each of
2 cells in a clone of embryonal carcinoma cells was impaled with a microelectrode. (b) As a current pulse (top trace) was injected into one cell, a
voltage deflection (bottom trace) was detected in that cell and also in the
other impaled cell (middle trace). The horizontal calibration bar is 100 msec,
and the vertical calibration bar is 1 nA.

adjacent cells but not of horse radish peroxidase (HRP) (M. W.
40,000). Cell injury which resulted in the loss of ionic coupling
also abolished the fluorescein and microperoxidase transfer. The
inability of injected HRP to move between cells via the gap
junctional channels allows its use as a diagnostic tool for
examining whether the presence of ionic coupling in any particular
case is mediated via junctional channels or via direct cytoplasmic
continuity. The presence of cytoplasmic continuity would result
in both ionic (or dye) coupling and the movement of injected HRP
between cells.

Gap junctions, as observed morphologically or inferred from
the presence of coupling, are found in almost all cell types that
have been examined, both in vertebrates and in invertebrates
(DeHaan & Sachs, 1972; Gilula, 1977; McNutt & Weinstein, 1973;

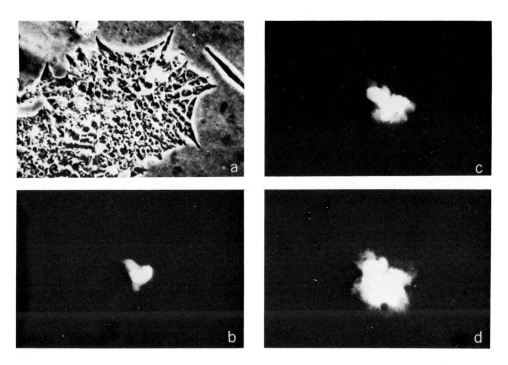

Figure 6: Transfer of injected fluorescein between embryonal carcinoma cells.
(a) Phase contrast image showing a clone of embryonal carcinoma cells in which
a single cell has been impaled with a fluorescein filled microelectrode.
(b-d) Fluorescein was injected into that cell and fluorescent images were
recorded at (b) 4 min, (c) 9 min, and (d) 17 min after the initiation of
injection.

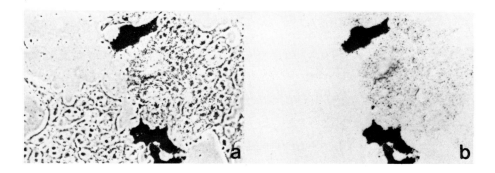

Figure 7: Metabolic cooperation between embryonal carcinoma cells. [H]³Uridine-
labeled embryonal carcinoma cells were co-cultured with unlabeled embryonal
carcinoma cells. After 4 hours, the cultures were fixed and processed for
autoradiography. (a) phase contrast image. (b) bright-field image. Note that
the stem cell clone at the bottom left which is not in contact with any radiolabeled cells has remained unlabeled.

Pitts, 1977; Satir & Gilula, 1973; Sheridan, 1974). The few cases
of exception are erythroid cells, skeletal muscle and most differentiated neuronal cells. It is not surprising that these cell
types do not express gap junctions since the presence of
junctional communication would be detrimental to the coordination
of muscular contraction and to the conductance of electrical
impulses by the neuronal elements. The absence of gap junctions
in erythroid cells would also be expected since they are circulating cells which rarely are involved in cell-cell, contact-
mediated interactions.

Most cells which communicate (as determined by ionic or
metabolic coupling) can communicate with any cell type, regardless
of the tissue or species of origin (Michalke & Loewenstein, 1971;
Pitts, 1972). However, in spite of the similarity of gap junctions
from cells of many different tissues and species, it seems that
the gap junctions of insects are somewhat different from the rest.
Morphologically, the gap junctions of insects (Gilula, 1977) are
distinct and can be considered as a separate class. In a study of
the communication competence of insect cells and mammalian cells
(Epstein & Gilula, 1977), it was observed that although the insect
cells are electrically coupled to one another and the mammalian
cells are also electrically coupled to one another, the heterologous combination of insect cells and mammalian cells is not
ionically coupled. The basis for this basic incompatibility of
mammalian and insect gap junctions is not known but it probably
involves structural differences in the gap junctions which have
accumulated in the years since the divergent evolution of insects
and mammals. In spite of junctional incompatibility, the
injection of fluorescent compounds of varying sizes and charges in
Chironomus salivary gland cells has revealed that the permeability
properties of insect gap junctions are similar to those of
mammalian gap junctions albeit perhaps somewhat less restrictive
(Flagg-Newton *et al*, 1979; Simpson *et al*, 1977).

3. GAP JUNCTIONS AND GRADIENTS

The prevalence of gap junctions throughout the animal kingdom
is somewhat confounding since their biological function is completely unknown. It has been suggested that gap junctional

communication may mediate the transfer of substances which
regulate and/or coordinate cell growth (Fentiman & Taylor-
Papadimitriou, 1977; Loewenstein, 1968, 1979; Pitts, 1978; Stoker,
1975) and tissue responses to hormonal stimulation (Lawrence et al,
1978; Pitts & Simms, 1977). It has been hypothesized that gap
junctions might in fact mediate the exchange of regulatory
molecules during embryogenesis (Furshpan & Potter, 1968;
Loewenstein, 1968; Pitts, 1977; Potter et al, 1966; Saxen et al,
1976); it is especially interesting to consider the possibility
that they might mediate the formation of morphogenetic gradients
(Wolpert, 1978).

Morphogenetic gradients can be defined loosely as gradients
of molecules which in some way provide information that leads to
the specific differentiation of cells in a position-dependent
manner (positional information), resulting in the formation of
predetermined patterns during embryogenesis (Cooke, 1975; McMahon
& West, 1976; Wolpert, 1969). The concept that the differentiation of cells during embryogenesis is specified by the position
of the cells or by positional information is based on the observations that sea urchin (Horstadius, 1973) and amphibian embryos
(Holtfreter & Hamburger, 1955; Spemann, 1938) have extensive
capacities for regulation. In such regulative embryos, the
developmental fate of the early blastomeres is relatively plastic
so that even if parts of the embryo are removed or transplanted
to a new site, normal development is still obtained. Thus the
potentiality of any cell is greater than what is normally expressed
during embryogenesis and the actual fate of any cell is determined
by its position relative to other cells within the embryo. The
specification of this positional information seems to occur in a
graded manner with defined polarities and thus has led to the
concept that gradients of some sort are responsible for specifying
the positional information. Diffusion (Crick, 1970; Munro &
Crick, 1970), active transport (Goodwin & McLaren, 1975), and
electrophoresis (Robinson & Jaffe, 1973) have all been proposed
as the motivating force in the erection of the gradient and models
of various types have been constructed to explore the possible
mechanisms whereby such presumptive gradients can be generated
(Gierer & Meinhardt, 1972; Goodwin & Cohen, 1969; Summerbell et al,

1973). Regardless of the mechanisms involved, gap junctions might provide an intracellular pathway via which such gradients can be formed in the developing embryo; extracellularly, the necessary concentration thresholds for establishing such gradients may be unable to form. The fact that gap junctions can indeed mediate the formation of gradients over long distances was clearly documented by Michalke (1977). Using quantitative autoradiography and metabolic cooperation as his assay, he observed that gradients of radiolabeled metabolites can be generated in a cell monolayer in which radiolabeled cells were present at only one end. In a matter of hours, a gradient over a distance of a millimetre or more was established. Therefore, provided that a "source" (and a "sink") is available, a gradient can be generated via junctional communication. Since gap junctions seem to function as passive molecular sieves with a finite pore size of 1,000 daltons or less, the molecules comprising such presumptive gradients must be rather small.

4. THE EARLY MOUSE EMBRYO

In the early mouse embryo, gap junctional communication as detected by ionic coupling and fluorescein transfer is first expressed at the 8-cell stage during early compaction (Figures 8 and 11) (Lo & Gilula, 1979a). The cell-cell transfer of injected HRP revealed that the presence of ionic coupling and fluorescein transfer detected earlier in the 2-, 4-, and the precompaction 8-cell stages is attributable to the presence of cytoplasmic bridges due to incomplete cytokinesis (Figures 9 and 10). These observations are in agreement with the electron microscopic studies in which gap junctions are observed in the 8-cell embryo (Ducibella & Anderson, 1975; Magnuson et al, 1977). Previously, in the developmental history of the oocyte, gap junction mediated communication was detected between the preovulatory oocyte and the follicular cells but after ovulation, communication was not detectable (Gilula et al, 1978). Interestingly, the first determination and differentiation event in the mouse embryo, that of trophoblast cell formation, occurs shortly after the onset of cell-cell communication at the late 8-cell stage. According to the hypothesis of Tarkowski and Wroblewska (1967), this first differentiation event in early mouse embryogenesis is regulated by

EARLY DEVELOPMENT OF MOUSE EMBRYO

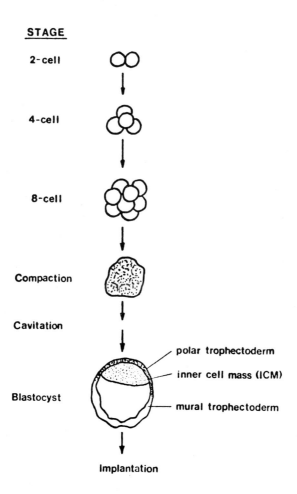

Figure 8: The development of the early mouse embryo. Cell-cell communication
is expressed de novo in the early compacted 8-cell embryo. Modulation of communication is observed in the post-implantation stages.

the position of the blastomeres in the embryo; thus only cells
located at the outside are triggered to differentiate into trophectoderm while the blastomeres in the inside form the cells of the
inner cell mass (ICM). Other experiments have shown that indeed
the position of the blastomere with respect to whether it is on
the outside or the inside of the embryo may be important in this

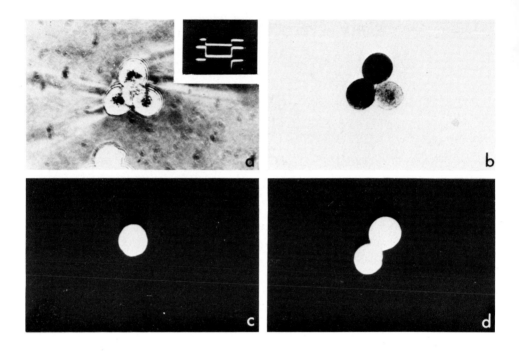

Figure 9: Ionic coupling in the 4-cell embryo via cytoplasmic bridges.
(a) Two cells of a 4-cell embryo were each impaled with a microelectrode.
(inset). As a current pulse (top trace) was injected into the upper impaled
cell, a voltage deflection was recorded in that cell (lower trace) and also
in the lower impaled cell (middle trace). A horizontal calibration bar =
100 msec and the vertical calibration bar = 5 nA, 20 mV for the middle trace
and 10 mV for the lower trace. (b) HRP was injected into the blastomere on the
left. The dark reaction product indicated the presence of HRP in two cells,
the cell into which HRP was injected and an adjacent blastomere.
(c,d) Fluorescein was injected into the left blastomere of the embryo and
fluorescence images were recorded at various times after the start of dye
injection. (c) 4 min, and (d) 14 min after the start of injection.

regulation of trophoblast versus ICM determination and differentiation (See Adamson & Gardner, 1979 for review; Gardner & Rossant,
1976; Rossant, 1978). The presence of gap junctions between the
cells of the embryo may perhaps allow the passage of metabolites,
ions and other molecules between the cells such that eventually
an intracellular gradient is generated which allows the distinction of the outside cells from the inside cells. In addition,
compaction is also accompanied by the formation of tight junctions
by the outside cells, leading to the generation of a permeability
seal which can further enhance the difference between the outside

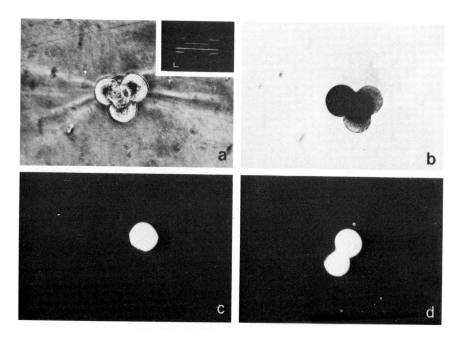

Figure 10: Communication in the 4-cell embryo via Cytoplasmic Bridges.
(a) Two cells of a 4-cell embryo were each impaled with a microelectrode.
(inset). As a current pulse (top trace) was injected into the impaled cell
on the right, a voltage deflection was detected in that cell (bottom trace)
but none was detected in the impaled blastomere on the left (middle trace).
The horizontal calibration bar = 100 msec and the vertical calibration bar =
5 nA and 20 mV. (b) HRP was injected into the impaled blasomere on the left.
The dark reaction product revealed the presence of HRP in two cells, the cell
into which HRP was injected and an adjacent blastomere. (c,d) Fluorescein
was injected into the blastomere on the right and fluorescence images were
recorded at various times after the start of dye injection.

versus inside microenvironment. This may be essential for
allowing the formation of a gradient of a critical steepness
necessary for eliciting an inside versus outside differentiation.
Thus the correlation in the timing of the onset of cell-cell
communication with the differentiation of trophectoderm and ICM
cells is consistent with the notion that gap junctions might
mediate the formation and transmission of a gradient of inside-
outside positional information.

        There are several intriguing observations made in early
mouse embryos which strongly suggest the importance of cell-cell

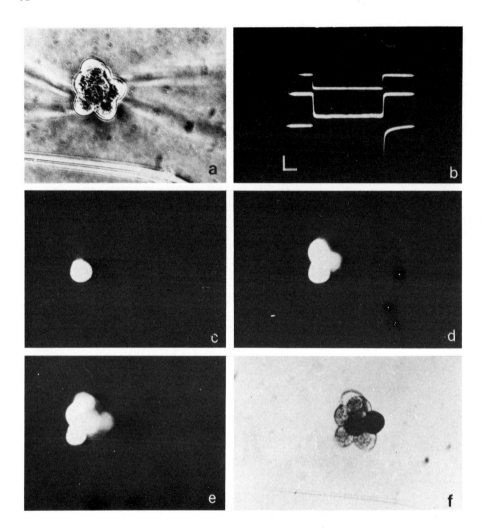

Figure 11: Junctional communication in the early compacted 8-cell embryo.
(a) Two cells of an early compaction 8-cell embryo were each impaled with a
microelectrode. (b) As current (top trace) was injected into the blastomere
on the left, a voltage deflection was detected in that cell (bottom trace)
and also in the impaled cell on the right (middle trace). The calibration
bars are the same as in Figure 10. (c-e) Fluorescein was injected into the
blastomere on the left and fluorescence images were recorded at various times
after the start of injection. (c) 4 min, (d) 16 min, and (e) 25 min after the
start of injection. (f) HRP was injected into the blastomere on the right.
The dark reaction product indicated the presence of HRP in two blastomeres;
the blastomere into which HRP was injected and also an adjacent blastomere.

contact in the regulation of early embryogenesis and are also consistent with the notion that gap junctions may mediate the passage
of regulatory informational cues. During early development,
shortly after compaction and as a result of tight junction
formation during compaction, the embryo forms a hollow cavitated
structure, the blastocyst (Figure 8). In the blastocyst, the
inner ball of cells or inner cell mass (ICM) gives rise to two
cell types, the primary ectoderm and the primary endoderm, and
the outer layer of cells forms trophectoderm. The cells of the
trophectoderm overlying the ICM, the polar trophectoderm, con-
tinue to undergo mitosis, while the trophectodermal cells away
from the ICM, the mural trophectoderm, undergo endoreplication
and form giant cells. As the polar trophectodermal cells con-
tinue to grow in number and become further removed from the ICM,
they also undergo endoreplication and give rise to giant cells.
Several different lines of experimentation suggest that the
intimate association of the ICM with the trophectoderm is responsible for maintaining the polar trophectoderm in a diploid and
mitotically active state and that without this cell-cell
association, the trophectodermal cells undergo endoreplication
(as in the mural trophectoderm), cease mitosis, and form giant
cells (see Gardner & Rossant, 1976, for review). Furthermore,
experiments by Pedersen and Spindle (1980) show that cell-cell
contact is important in the regulation of trophectodermal and
ICM differentiation. They microinjected 8-cell and morula stage
embryos into the blastocoel of a giant chimaeric blastocyst
(formed by the aggregation of 8-10 embryos during the 4- to 8-cell
stages) and found that if the zona was not removed from the donor
embryos, then normal blastocysts were formed in most cases.
However, when the zona was removed prior to injection, the donor
embryos often formed a compacted ICM-like structure attached to
the trophectodermal wall of the recipient blastocyst. Hence it
seems that cell-cell contact is critical in the regulation of
differentiation of the totipotent blastomeres and that the blastocoelic fluid does not either inhibit trophoblast differentiation
or promote ICM differentiation. In addition, studies by Johnson
et al (1979) and Kemler et al. (1977) have demonstrated that
inhibition of compaction in early mouse embryos (as a result of

exposure to antisera raised against mouse embryonal carcinoma cells) resulted in the reversible arrest of development such that if the antisera were removed subsequently, the embryo resumed compaction and normal development. In this case, although cell-cell contact was maintained, development was arrested nevertheless. Such observations thus suggest that cell-cell contact alone is insufficient and perhaps the formation of specialized junctions is necessary for normal embryogenesis to proceed. It would be interesting to know whether gap junctional communication is still present in such embryos and ultimately, in order to discern the role of cell-cell communication in embryogenesis, it would be necessary to be able to specifically inhibit gap junctional communication and study the subsequent effects on development. Since at the present, there are no known means of specifically inhibiting gap junctional communication, this type of antisera Johnson *et al*, 1979; Kemler *et al*, 1977) may provide an indirect means for probing this question.

5. OTHER EARLY EMBRYOS

The cell-cell communication properties of other early embryos have also been studied and ionic coupling has been observed in most instances. Ionic coupling was found in the starfish (Ashman *et al*, 1964; Tupper *et al*, 1970; Tupper & Saunders, 1972), the squid (Potter *et al*, 1966), the fish, *Fundulus* (Bennett & Trinkaus, 1970; Bennett *et al*, 1972), the newt, *Triturus* (Ito & Hori, 1966; Ito & Loewenstein, 1969) the South African clawed toad, *Xenopus* (DiCaprio *et al*, 1975; Sheridan, 1971; Slack & Palmer, 1969), the frog, *Rana pipiens* (Woodward, 1968), the chick (Sheridan, 1968), and the lobster (Furshpan & Potter, 1968) embryos. The possible presence of cytoplasmic bridges due to incomplete cytokinesis (which often exists during early cleavages in embryos) was not determined in these studies, therefore it is not possible to unequivocally attribute the ionic coupling detected by these workers to gap junction mediated coupling. The cells of the amphibian embryo are all apparently coupled from the 2-cell stage on. In the starfish embryo, ionic coupling was not detected until the 32-cell stage, and at that time the passage of injected fluorescein was not observed (Tupper & Saunder, 1972). Similarly, fluorescein injections of early *Xenopus* embryos (Slack & Palmer, 1969) and *Fundulus* blastomeres (Bennett *et al*, 1972) revealed the apparent inability of the injected dye to pass between the embryonic cells although ionic coupling was present. These observations led to the suggestion that gap junctions of embryonic cells may be different from the gap junctions of adult cells in that either the pore size is smaller or the channels may be selective so that only certain types of molecules can be exchanged.

These observations in the embryos are, however, contradictory to most other studies of cell-cell communication in which, consistent with the known exclusion limit of junctional channels, it has been found that if a cell is competent to undergo ionic coupling, it can also pass injected dye and/or undergo metabolic coupling (Figures 5, 6 and 7) and that when coupling is abolished, the ability to transfer injected dye is similarly lost. Thus, the apparent anomaly observed with embryonic cells was rather exciting, because it suggested that selective permeability or reduced pore size might be the clue to some important functional role for gap junctions in early embryogenesis. However, further analysis of dye coupling in *Fundulus* embryonic cells indicated that the previous failure to observe dye coupling was more apparent than real and that in fact, under more careful scrutiny, the passage of injected fluorescent tracers was observed (Bennett *et al*, 1978). Therefore, the failure to observe dye passage in *Asterias* and *Xenopus* embryos must also necessarily be carefully re-examined. It is possible that in embryos where yolk granules occupy a large part of the cytoplasm of early blastomeres (as in *Xenopus*, *Fundulus* and *Asterias*) the passage of fluorescein or other tracer dyes may be difficult to visualize. In addition, the cell volume of such early blastomeres is rather large so that a large amount of injected dye may be needed before it becomes readily visible against the background. For these reasons, unless a sufficient amount of dye is injected, the passage of dye between cells may often be missed. For mouse embryos, in which the blastomeres are rather small (in the early 8-cell stage, the blastomeres are less than 20 µm in diameter) and the content of yolk granules is minimal, the presence or absence of dye passage

is readily discernable. Therefore, at the present time, there
is no strong evidence to support the notion that gap junctions of
embryonic cells are different from those of other cell types.

6. ABSENCE OF CELL-CELL COMMUNICATION

In light of the fact that most embryonic cells have gap
junctions and are therefore probably competent in communication,
it has been suggested that the study of cases in which communication does not exist are perhaps more relevant to understanding
the role of gap junctions in embryogenesis. There are several
known examples in which cell-cell communication as monitored by
ionic coupling was found to be absent at specific times during
embryonic development (Blackshaw & Warner, 1976; Dixon & Cronly-
Dillon, 1972; LoPresti et al, 1974; Potter et al, 1966; Warner,
1973). For example, in the first study of ionic coupling in
embryos, Potter et al. (1966) observed that the cells of the
squid embryo were initially all coupled to the yolk cell but that
as development progressed, this coupling was completely lost.
In Bombina and Ambystoma, the cells of the myotome and dermatome
layers lose their ionic coupling with one another once the somites
have formed (Blackshaw & Warner, 1976) and in the axolotl, the
ectodermal cells become uncoupled from the neural cells during
the time of neural tube closure (Warner, 1973). The significance
of such breaks in cell-cell communication is not known and there
are no a priori reasons to suspect the necessity of such breaks
in communication in these instances. However, in a study of the
developing retina of Xenopus (Dixon & Cronly-Dillon, 1972) gap
junctions were observed to disappear from the central retina at
the time of the specification of retinal polarity. This correlation in timing between gap junction removal and the onset of
the determination of polarity is consistent with the notion that
the positional information required for directing specific dif-
ferentiation/determination can pass between cells via junctional
pathways. In the case of the insect cuticle, specific breaks
in cell-cell communication would also be predicted if such a
notion were valid. The pattern of insect cuticle is specified
by the underlying epidermal cells which are responsible for its
biogenesis. A comparison of the cuticular pattern in the normal
insect with those in which the integument has been transplanted or

rotated (Locke, 1959, 1960, 1967; Lawrence, 1972) revealed that
within each segment, there is a gradient of positional information
which determines the polarity and the specific differentiation of
the epidermal cells and that the gradient, whose high and low
points are defined by the segmental boundary, is repeated within
each segment. In addition, cell lineage studies revealed that the
segmental boundary represents a "compartment" border so that
clones generated after the blastoderm stage never cross the segmental boundary (Lawrence , 1973).   Thus each segment is derived
from a small group of primordial cells ("polyclones") which once
having been set aside for one segment (compartment) will never
cross over to an adjacent segment (compartment). If gap junctions
were involved in mediating the transmission of morphogenetic
gradients, which presumably would direct the differentiation
within each compartment, then cell-cell communication ought not
to be present between the epidermal cells of two adjacent segments
at the segmental boundary. Ultrastructural analysis demonstrated
the presence of gap junctions between cells at the boundary
(Lawrence & Green, 1975) and, moreover, studies by Warner and
Lawrence (1973) demonstrated the existence of ionic coupling
across the segmental boundary. Such findings are extremely
provocative and suggest that either gap junctions are not involved
in mediating the formation of presumptive morphogenetic gradients
or that cell-cell communication at the segmental boundary must
somehow be different from the cell-cell communication within each
segment. Such hypothetical differences or discontinuity in the
communication properties of the cells at the segmental boundary
would be essential to the formation of independent morphogenetic
gradients (via gap junctional channels) within each developmental
unit.

7.   MODULATION OF CELL-CELL COMMUNICATION

7.1 *CELL SHAPE*

In spite of the fact that all gap junctions are probably
very similar, the ability of cells to communicate with one another
may still be modulated by other parameters. This is a particularly
important point since without a means of modulating communication,
gap junctions would be a much less interesting proposition for

58

developmental biologists. This is due to the fact that most
embryonic cells studied are found to be linked by gap junctions
and if cell-cell communication were not modulated, most of the
cells of the embryo would therefore behave as a massive syncytium.
Thus there would be no basis from which separate gradients can
be generated and developmental compartments formed. This is, of
course, clearly illustrated in the case of the insect segmental
boundary. Given that all gap junctions are the same, that is
they all have the same passive sieve-like property with the same
exclusion limit, then the most obvious manner in which cell-call
communication can be modulated is through modulating the number
of gap junctions present between cells. Therefore, given that
all else is equal, the total number of gap junctions present at
any one time would determine quantitatively the rate of exchange
between cells. The lower and upper limits of this exchange are,
respectively, the complete absence of exchange where no gap
junctions are present, and rapid exchange in which diffusion
becomes the only limiting step.

In studies of metabolic cooperation between fibroblasts
and epithelial cells, Fentiman et al. (1976) and Pitts & Burk
(1976) found that even though the fibroblast or epithelial cells
cooperated very well amongst themselves, they were very poor in
metabolic cooperation with one another, i.e. the heterologous
combination of epithelial cells and fibroblast cells did not
exchange tritiated nucleotides as efficiently as the homologous
combination of epithelial cells or fibroblast cells. They considered this poor exchange to result from the inability of the
two cell types to make efficient contact with one another,
probably as a result of the drastic differences in the cell
shapes of fibroblasts versus epithelial cells. This seems a
very plausible explanation since epithelial cells are usually
rather small cells with a large nuclear to cytoplasmic ratio and
grow in rather tight "nests" while fibroblasts which have a low
nuclear to cytoplasmic ratio are rather large and are usually
spread out. This striking phenomenon was also observed with
mouse embryonal carcinoma cells (mouse teratocarcinoma stem
cells), also an epithelial-like cell type (Lo & Gilula, 1980b). In

co-culture with a variety of cell types, embryonal carcinoma cells were very poor recipients in metabolic cooperation except when co-cultured with themselves or BRL (rat liver cell line) which is also an epithelial cell type (Table 1). In contrast, these embryonal carcinoma cells were very good donors in metabolic cooperation with most of these same combinations of various cell

TABLE 1: METABOLIC COUPLING BETWEEN VARIOUS CELL TYPES*

	Recipient									
CELL TYPES	STO	3T3	MF	CF	CHO	PCC4azal (STEM)	BRL	MDCK	PYS	PCC4azal (GIANT)
DONOR STO	+	+	+	+			+	+		
3T3	+	+	+	+		−	+	+	+	+
MF	+	+	+	+	−	−	+	+		
CF	+	+	+	+		−	+	+		
CHO	+	+	+	+	+	−	+	+		
PCC4azal (STEM)	+	+	+	−	−	+	+	−	+	+
BRL	+	+	+	+	+	+	+	+	+	+
MDCK	+	+	+	+	+	−	+	+	+	+
PYS		+					+	+		

* The degree of metabolic coupling between the various cell types was scored as (+) for high incidence, and (−) for very low incidence of transfer of ^{3}H-uridine from donor cells to recipient cells. No attempt was made to quantitate the data with grain counts since the qualitative differences noted in the above table were readily apparent.

types. This discrepancy may be most easily explained by the clonal morphology of the embryonal carcinoma cells at the time of the experiment, i.e. the donor stem cells were plated as single cells onto unlabeled recipient cells while the recipient stem cells were in tightly packed epithelial clones when co-cultured with radiolabeled donor cells. The tight packing of the embryonal carcinoma cells probably greatly limits the extent of cell-cell contact with heterologous cell types. This in turn limits the

number or rate of gap junctions formed and hence reduces the
amount of metabolic coupling observed between the embryonal
carcinoma cells and other donor cell types which cannot be easily
integrated into the epithelial nests of the embryonal carcinoma
cells. These observations could not be simply explained by any
differences in either the nucleotide pool sizes or the rate of
incorporation of tritiated uridine since the embryonal carcinoma
cells are good donors in the same heterologous cell combinations
and all the heterologous cell types tested are both good donors
and recipients in homologous and heterologous combinations with
each other (Table 1). These findings are especially interesting
in the light of the fact that embryonal carcinoma cells may be
very similar to the early stem cells of the mouse embryo (Stevens,
1967; Martin, 1975) and may be equivalent to the cells of the
ICM, that is, they both have a wide range of differentiation
potential, grow in tightly packed nests, and are similar in
ultrastructural morphology. Perhaps the specificity in the
efficiency of embryonal carcinoma cells to communicate with other
cell types may reflect the importance of cell-cell communication
in the regulation of the differentiation and development of the
cells of the ICM. In the case of the embryonal carcinoma cells,
it has been shown that when injected into a mouse blastocyst,
some cells may convert from a tumorigenic cell type to a normal
embryonic phenotype and differentiate as normal totipotent
embryonic stem cells (Mintz & Illmensee, 1975). It is conceivable
that upon injection into the mouse blastocyst, they are readily
incorporated exclusively into the epithelial nests of the ICM
while rapidly establishing effective communication with the ICM
cells and as a result are converted to a normal embryonic phenotype. That the blastocoelic fluid may actually have the crucial
influence on the embryonal carcinoma cells rather than any cell-
cell interaction mediated effect is possible but, in the light of
the findings of Pedersen and Spindle (1980), seems less likely.

This type of specificity in cell-cell communication is not
exhibited by all epithelial or fibroblast cells (Fentiman et al,
1976; Lo & Gilula, 1980b; Pitts, 1978). For example, bovine lens
epithelial cells are equally good at metabolic cooperation with
other epithelial cells as well as with fibroblast cells (Fentiman

et al, 1976). Perhaps the explanation for these differences lies
in the fact that the cell shape and size of what are considered
as epithelial cells or fibroblast cells may actually vary greatly
from one cell line to another; sometimes there is even great
variation in cell shape and size within a single clonally-derived
cell line. Thus, what may be important in specifying the
efficiency of cell-cell coupling is not whether the cells are
fibroblasts or epithelial cells but rather whether the overall
cell morphology of the various cell types would allow the formation of close cell-cell contact.

Cell shape may therefore provide a means whereby the extent
of cell-cell communication in intact tissues of embryos and adults
may be modulated quantitatively through the modulation of the
number of junctional channels formed between various cell types.

7.2 BASEMENT MEMBRANE

Cell-cell communication may also be modulated via the
formation of basement membrane, which can prevent cell-cell contact
and hence potentially prevent the formation of gap junctions
between cells. Of course, whether such a scheme can be used *in
vivo* to control specifically gap junctional communication or
whether the stoppage of communication would only be an unimportant
consequence of the formation of the basement membrane remains to
be established. Moreover, transient discontinuities in the basement membrane may certainly exist as a result of the constant
remodeling of the basement membrane in developing tissues and it
is not clear whether cell-cell contacts and gap junctions can be
made through such transient breaks. However, it is necessary to
keep in mind the possibility that in the modulation of cell-cell
communication, what may be important in some cases is not the
absolute presence or absence of communication but rather the fine
tuning of how much, where, and when.

7.3 TIGHT JUNCTIONS

Another means whereby cell-cell communication may be controlled is through the formation of tight junctions. This was
suggested by studies probing the metabolic cooperation competence
of epithelial cells which form tight junctions *in vitro*.
Epithelial cells such as MDCK (dog kidney epithelial cells)

(Cereijido *et al*, 1973), BRL (rat liver epithelial cells), or the differentiated endoderm-like teratocarcinoma cells (Lo & Gilula, 1978a) will, upon growth to confluence *in vitro* form a cell monolayer which is connected by a network of tight junctions at their apical cell surfaces. Such epithelial structures formed *in vitro* resemble a typical epithelium *in vivo* with a distinct polar arrangement of the cell monolayer consisting of an apical and a basal side; the extensive tight junctional complexes that are formed give rise to a permeability seal similar to the intact epithelium so that diffusion from the apical to the basal side is limited. BRL and MDCK cells plated at low density do not form such tight junctional networks and can metabolically cooperate with any other homologous or heterologous cell type. However, at confluence, when tight junctions are formed, these cells cannot undergo metabolic cooperation with any radiolabeled donor cells, either homologous or heterologous (Lo, unpublished observation). Thus, the apical surface of the epithelial monolayer which contains a complete tight junctional network is impenetrable with regard to gap junction formation and cell-cell communication. Perhaps *in vivo* this can serve not only as a barrier to isolate cells by preventing diffusion but also as a barrier which can further isolate cells by preventing cell-cell communication.

Several different lines of studies have suggested yet other means of modulating cell-cell communication in which the actual number of gap junctional channels does not change; rather, control seems to occur by specifically opening and closing a certain number of the existing channels.

7.4 CALCIUM

Calcium has been implicated as playing such a regulatory role in the *Chironomus* salivary gland cells (Deleze & Loewenstein, 1976; Loewenstein, 1966; Oliveira-Castro & Loewenstein, 1979; Rose & Loewenstein, 1975; Rose & Loewenstein, 1976). It has been demonstrated that the permeability of the junctional channels depends on the cytoplasmic calcium concentration. At normal calcium concentrations ($<10^{-7}$ M) the junctional channels are highly permeable (Simpson *et al*, 1977), at high calcium concentrations ($>5 \times 10^{-5}$ M), the channels are impervious to both ions and low molecular weight dyes such as fluorescein (Rose &

Loewenstein, 1976; Oliveira-Castro & Loewenstein, 1971), and at intermediate calcium concentrations ($<5 \times 10^{-5}$ M) the transfer of fluorescein is reduced while the passage of ions is completely blocked (Deleze & Loewenstein, 1976). Furthermore, Rose and Loewenstein (1977) recently reported that the elevation of intracellular calcium concentration can produce "graded" changes in permeability such that the larger of two fluorescent molecules being coinjected is retarded in cell-cell transfer while the transfer of the smaller molecular species is unaffected. Taken together, these observations seem to suggest that calcium can close junctional channels and that the closing involves a graded drop in channel pore size. Thus it was suggested that by regulating the level of intracellular calcium, the extent of cell-cell communication can be modulated by selective restriction of the junctional channel permeability. There are several still unresolved issues in these studies, such as what is the role of membrane depolarization, which usually accompanies the calcium mediated uncoupling, in the uncoupling process itself and how accurate are the aequorin-dependent determinations of intracellular calcium concentration (see Sheridan, 1978).

7.5 INTRACELLULAR pH

More recently, other studies have shown that a decrease in intracellular pH can also cause a decrease in junctional conductance (Bennett et al, 1978; Hanna et al, 1978; Turin & Warner, 1977). It has been suggested that the drop in junctional conductance observed with calcium injections may actually result from a decrease in intracellular pH caused by the compartmentalization of calcium in different regions of the cytoplasm. It remains to be seen whether such decreases in pH or elevation in calcium concentration in the cytoplasm are physiologically feasible and relevant.

7.6 TRANSJUNCTIONAL VOLTAGE

In another study (Spray et al, 1979), it was found that transjunctional polarization can also cause a decrease in junctional conductance. Thus, by altering the resting membrane potentials it may be possible to regulate the level of junctional communication. The development of large differences in resting

membrane potentials between different regions of the embryo has
been documented (Blackshaw & Warner, 1976) which can conceivably
serve as a mechanism for the specific modulation of the extent of
cell-cell communication during embryogenesis.

These observations all suggest the possibility that cell-cell
communication, at least in some cells, can be modulated by
directly regulating the permeability of the junctional channels
themselves. The applicability of these findings to mammalian
cells and whether these mechanisms are normally operative in the
developing embryo remain to be examined.

8. MODULATION OF GAP JUNCTIONAL COMMUNICATION IN THE POST-IMPLANTATION MOUSE EMBRYO

In examining the cell-cell communication properties of the
postimplantation mouse embryos *in vitro*, some surprising
observations were made which suggest that indeed there is modulation of gap junctional communication in the intact developing
mouse embryo (Lo & Gilula, 1979a,b). The mouse embryo forms a
blastocyst on day 4 of development and subsequently will undergo
'implantation' *in vitro*. This process is initiated *in vivo* by
the invasion and infiltration of the uterine epithelium by the
trophoblast cells and results in the firm attachment of the
developing embryo to the uterus, culminating in the formation of
the placenta. Blastocysts which are cultured *in vitro* will mimic
the implantation process by spreading out on and attaching firmly
to the surface of the culture vessel. In this implantation-like
process, the trophoblast cells migrate outward, forming a ring of
fibroblast-like cells surrounding a central mass of small cells,
the cells of the ICM (Figure 12). The trophectodermal cells,
which are initially relatively small and mononucleated, undergo
endoreplication in a period of 2-4 days, giving rise to giant
cells which often are also multinucleated (Figure 13). The cells
of the ICM similarly undergo further development upon implantation
in vitro; they increase in number and differentiate into two
distinct cell types, the ectoderm and the endoderm (Figure 13).
These developmental processes closely parallel the sequence of
events in the implanted embryo *in utero*, although the rate is
somewhat slower. If a great deal of care is exercised with regard
to the culture conditions, such embryos can develop up to the

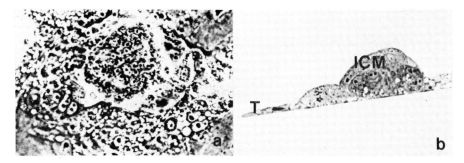

Figure 12: The *in vitro* implanted mouse embryo shortly after attachment.
(a) Phase-light micrograph. The central group of small cells which contain
large nuclei with prominent nucleoli are the cells of the inner cell mass.
The surrounding cells are trophectodermal and at this stage their cytoplasm
is filled with refractile vacuoles. (b) Thick-section at a perpendicular plane
through *in vitro* implanted mouse embryo similar to the one in (a). The large
central mass is the inner cell mass (ICM) and the flat cells in the periphery
are the trophectoderm (T).

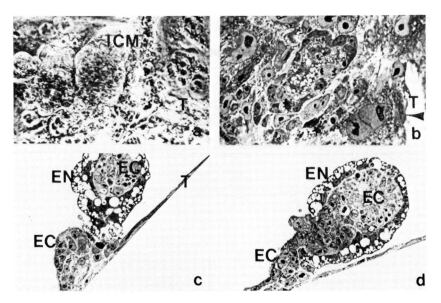

Figure 13: The *in vitro* implanted embryo at a later stage. (a) Phase micrograph
of a late attached mouse embryo. The 3-dimensional mass in the centre is the
inner cell mass (ICM). The large flat cells (T) in the periphery are trophoblastic. (b) Thick section through the horizontal plane of such an embryo. The cells
in the centre with small nuclei are the cells of the inner cell mass. The
surrounding cells with large nuclei are the trophoblast (T) or giant cells.
(c) and (d) Two thick sections cut through a perpendicular plane of such an
embryo. The large cell mass is the ICM; it consists of two morphologically distinct cell types, the ectoderm (EC) and the endoderm (EN). The cytoplasm of endoderm cells is filled with large vacuoles. The flat cells (T) are trophoblastic.

twelfth somite stage (Hsu, 1979). This system would thus allow the use of microelectrodes to directly examine the cell-cell communication properties of the intact and normally developing postimplantation mouse embryo. The extent of communication between different regions of the embryo was analyzed by probing for the presence of ionic coupling and the extent of dye spread after microinjection with the fluorescent dyes, fluorescein or Lucifer yellow. In the early *in vitro* implanted embryo, all the cells of the embryo are ionically coupled, i.e. the ICM cells are coupled to each other as are the trophoblast cells and in addition the ICM cells and the trophoblast cells are also coupled to one another. This was further confirmed by the fluorescein injections which revealed the complete spread of the injected dye throughout the entire embryo (Figure 14), regardless of whether the injection site was a trophoblast cell or an ICM cell. As development ensued (including trophoblast giant cell formation and differentiation of ectoderm and endoderm) examination for ionic coupling and dye spread revealed that the dye spread became more and more limited until it was confined to within sharp boundaries while ionic coupling was maintained throughout the entire embryo. Thus injection of dye into the ICM region resulted in dye spread throughout the ICM but less and less dye spread into the surrounding trophoblast cells, until in the later stage embryos, no dye spread to the trophoblast cells was observed at all (Figure 15). In these later stage embryos, injection of dye into a trophoblast cell (which had become a giant cell) resulted in the filling of that cell only, with no dye spread to any other cell (Figure 16). Moreover, dye injection into the ICM region in the later stage embryos resulted in only limited spread to certain groups of cells within the ICM region while other regions of the ICM remained completely void of the injected dye (Figure 17). This sequential segregation of dye spread into the embryo seems to roughly parallel, chronologically and geographically, the for-mation of trophoblast giant cells and the differentiation of the ICM cells into endoderm and ectoderm cells. The most intriguing observation in these studies is the consistent finding that throughout these various stages in which the dye-spread pattern became more and more limited, ionic coupling was maintained

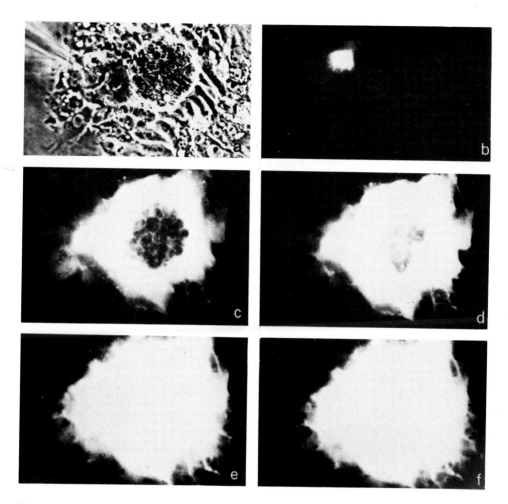

Figure 14: Spread of injected fluorescein in the early attached mouse embryo. (a) A trophoblast cell of an early attached mouse embryo was impaled with a microelectrode. (b-f) Fluorescein was injected into the trophoblast cell and the fluorescence images were recorded at various times after the start of injection: (b) 4 min, (c) 8 min, (d) 16 min, (e) 20 min, and (f) 31 min.

across the boundary beyond which no dye spread was observed (Figure 15). Thus, cells which are not coupled with respect to dye transfer are, nevertheless, ionically coupled. These observations can be explained in several ways: that there are two types of gap junctions with different pore sizes, that the pore size of the gap junctional channels can be selectively regulated,

or that the ionic coupling observed is not mediated by gap
junctions but due to the presence of tight junctions.

The presence of tight junctions was monitored by examining
freeze fracture replicas of embryos implanted *in vitro*. By the
early implantation stage, most of the tight junctional elements
(which were associated with the trophoblast cells) were no longer
present. Therefore, tight junctions could not have been responsible for the observed ionic coupling in the later stage embryo.
The existence of gap junctions with very different pore sizes
is also an unlikely possibility since thus far there is no
evidence from any study to support this idea; rather, the evidence
would tend to suggest that normally most gap junctions resemble
one another. The third possibility that the junctional channels
may be selectively permeable is an intriguing notion but, aside
from observations in the *Chironomus* salivary glands, there is no
evidence that such a mechanism may be operative in mammalian
cells or in any other cell type.

9. MODULATION OF CELL-CELL COMMUNICATION BY REDUCTION IN FLUX

A simpler interpretation of the presence of ionic coupling
without dye transfer in cells of the postimplantation mouse embryo
is that the lack of dye transfer is not the result of dye molecules being unable to pass, but only that the number of dye molecules that are passed is below a threshold level that is required
for visibility; whereas ionic coupling, being a more sensitive
assay of junctional communication, can detect the passage of ions
even though the passage of injected fluorescein is not observable.
Thus the detection of ionic coupling without dye transfer may in
fact reflect differences in flux (rate of molecules transferred/
area of junctional membrane) at the boundary. Such a change in
flux may simply result from a decrease in the number of functional channels interconnecting cells at the boundary regions.
Accordingly, in the late embryo, fluorescein injected into the
ICM may still be transferred to the trophoblast but at such a slow
rate that the rate of fluorescence quenching exceeds the rate at
which the dye accumulates, resulting in the apparent lack of
visible dye transfer. The absence of dye transfer between different regions of the ICM and between the giant trophoblast cells

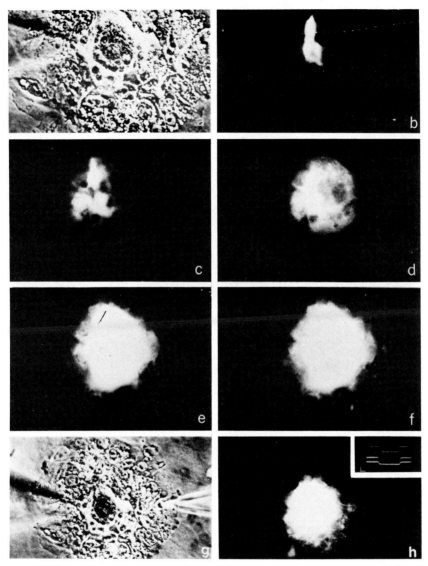

Figure 15: Ionic coupling across the boundary of dye-spread in a mouse embryo
at a later stage of development. (a) The embryo was impaled with microelectrodes in a trophoblast cell and in a cell bordering on the ICM region.
(b-f) Fluorescein was injected into the cell in the ICM region and the fluorescence images were recorded at (b) 4 min, (c) 9 min, (d) 12 min, (e) 19 min,
and (f) 31 min after start of injection. (g-h) Lower mag. of (a) and (f). Note
that the impaled trophoblast cell is beyond the region filled by the dye.
(Inset) As a current pulse (top trace) was passed during the fluorescein
injection from the microelectrode on the left, a voltage deflection was detected
in that cell (middle trace) and also in the trophoblast cell (bottom trace) on
the right. The horizontal calibration bar = 100 msec and the vertical calibration bar = 5 nA and 10 mV.

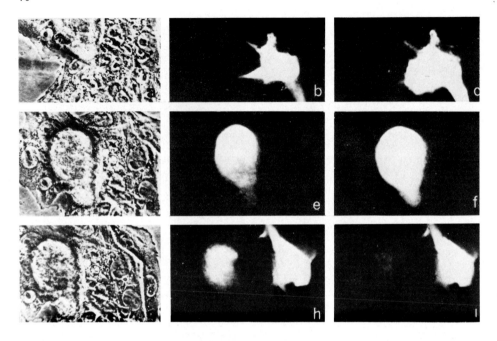

Figure 16: Spread of injected fluorescein in the mouse embryo at a later stage. (a, d, g) Well developed embryo was successively impaled with a microelectrode in three different regions: (a) a giant trophoblast, (d) inner cell mass, and (g) another giant trophoblast. (b-c, e-f, h-i) Fluorescein was injected into the impaled cells and the fluorescence images were recorded at various times after the start of injection: (b) 4 min, (c) 22 min, (e) 4 min, (f) 28 min, (h) 4 min, and (i) 19 min.

may reflect similar changes in flux due to a decrease in the number of functional channels. Such a hypothetical decrease in the number of functional channels can be achieved very simply by an actual reduction in the number of gap junctions interlinking cells at the boundary region. This type of modulation of com- munication may resemble what was observed in the modulation of metabolic cooperation as discussed previously, in which it was thought that the presence of a smaller number of gap junctions resulted in a lower amount of nucleotide exchange being observed. In the previous case, differences in cell shape were thought to contribute to the poor metabolic coupling but the cause for the aberration in dye coupling in the mouse embryo is not so obvious. Whatever the exact cause(s) may be, cell shape alone cannot account for all of the observations in the mouse embryo since the trophoblast cells of the later stage embryo (i.e. the giant cells)

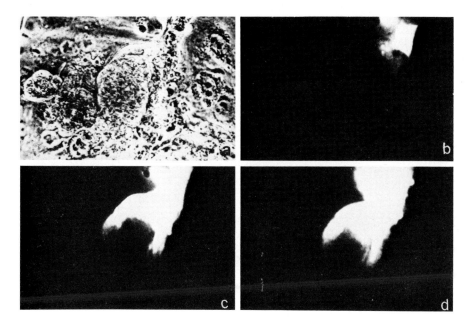

Figure 17: Segreation of fluorescein dye transfer in the ICM of a well developed mouse embryo at a late stage. (a) A cell in the ICM region of the mouse embryo was impaled with a fluorescein-filled microelectrode. (b-d) Fluorescein was injected into the impaled cell and fluorescence images were recorded at various times after the start of dye injection: (b) 4 min, (c) 13 min, and (d) 31 min.

which presumably are still ionically coupled to one another, do not pass injected fluorescein even though they have maintained their fibroblast morphology.

The reduction in the number of functional channels may also result from the closing of existing channels without necessitating an actual reduction in the number of gap junctions. The observations of reduced ionic coupling as a result of either decreases in intracellular pH or increase in transjunctional membrane potentials both suggest means whereby the extent of communication can be quantitatively regulated by the closing of junctional channels. Therefore the finding of limited dye transfer in the mouse embryo can readily be explained by a reduction in the number of functional junctional channels (by whatever means) between the trophoblast cells and between the trophoblast cells and the ICM cells at the boundary regions.

Studies of the metabolic cooperation in mouse embryonic cells (Gaunt & Papaioannou, 1979; Lo, unpublished observations) have also indicated that trophoblast cells of the *in vitro* implanted embryo cannot undergo coupling with other cell types; this is consistent with the observation of limited dye spread. Again, the lack of observable metabolic coupling could presumably arise from a low level of junctional communication, resulting in few radiolabeled nucleotides being fixed into nucleic acids and hence making any nucleotide transfer not observable in autoradiography. The limitation here again is visual, the detection of specific incorporation against background grains in autoradiography; other variables which will also determine the detectability of metabolic coupling include the size of the nucleotide pool, the rate of nucleotide incorporation, and the rate of their turnover.

10. COMMUNICATION COMPARTMENTS AND GRADIENTS

Taken together, the observations summarized above indicate clearly that in the early mouse embryo, modulation of cell-cell communication does occur and results in the formation of semi-independent communication compartments: the dye spread pattern delineates the separate communication compartments while the ionic coupling demonstrates the persistent but low level of communication between compartments, thus making them semi-independent compartments. This partial segregation of cell-cell communication in the postimplantation embryo roughly parallels specific differentiation processes, in particular that of trophoblast giant cell formation and ectoderm and endoderm differentiation. Were gap junctional communication involved in the establishment of morphogenetic gradients, then such communication compartments would be necessary to allow both the generation of a concentration threshold of the molecules of each gradient and the formation of such gradients in each compartment independently. The timing in the breakoff in dye spread observed between the trophoblast and the ICM is consistent with such a hypothesis. Normally, after implantation *in utero*, the embryo becomes closely associated with the maternal uterine epithelial cells as a result of the trophoblast cell invasion and gap junctions are formed between the embryonic trophoblast cells and the maternal uterine cells

(Tachi *et al*, 1970). If morphogenetic gradients of the ICM are
forming via junctional channels, then it may be critical to
isolate this communication compartment from the maternal cells
at the time when trophoblast cells are making junctional contacts
with the uterine cells; otherwise gradients of the appropriate
steepness and the required geometrical distribution may be unable
to form. Moreover, the further segregation of dye spread within
the ICM of later stage embryos may reflect the formation of other
communication compartments that are necessary for the generation
of morphogenetic gradients involved in the further differentiation
of the embryonic ectoderm within the ICM. The low level of
junctional communication which continues to connect neighbouring
communication compartments may be important for coordinating the
differentiation events between several different compartments
without affecting the formation of separate morphogenetic
gradients. In this manner, regulative control between developmental compartments can be exercised in the context of the
establishment of semi-independent communication and developmental
compartments. Thus during development, the formation of (semi-
independent) communication compartments can be envisioned to
occur many times, even within existing compartments, each time
allowing the generation of morphogenetic gradients which then
elicit the appropriate determination/differentiation events.

The finding of ionic coupling between the epidermal cells
from neighbouring segments in *Rhodnius* as discussed previously is
perhaps another example of the presence of semi-independent communication compartments, as in the mouse embryo. Whether this is
the case would require further study to examine the extent of
spread of injected tracer dyes across segmental boundaries.

This hypothesis of gap junction mediated formation of morphogenetic gradients can also explain the fascinating observation of
polyclones in insect development (Crick & Lawrence, 1975; Garcia-
Bellido, 1975), in which the cell lineage of any adult structure
can only be traced back to a group of cells in the larvae, never
to just one cell. In the development of the mouse, similar
observations were made with the germ cells, the follicle cells,
and the pigmented cells of the coat and the retina. Using
chimaeric embryos and the resulting allophenic mice, it was found

that all of these cell types can actually be traced back to a fixed number of primordial cells (Mintz, 1967, 1971). These observations may be explained simply by the supposition that only when a critical number of cells are present can a communication compartment of the required size and geometry be formed, and only then can the necessary morphogenetic gradients be generated to elicit the appropriate determination/differentiation events.

11. CONCLUSION

It is hoped that in future studies, as more is learned about cell-cell communication in other early embryos, a more thorough evaluation of the possible role of gap junctions in development can be made. The existence of communication compartments which also delineate known developmental compartments would strongly implicate gap junctional communication as the means whereby morphogenetic gradients are generated. This perhaps should be the immediate challenge for those investigators who are eager to find more conclusive (but still indirect) evidence for the involvement of gap junctions in embryogenesis. Ultimately, this question can be directly approached only when the components of such hypothetical gradients are identified and a method becomes available for specifically inhibiting junctional communication.

ACKNOWLEDGMENT

The author would like to thank N. B. Gilula, in whose laboratory much of this work was carried out; Rocky Tuan, for reviewing and assistance in preparing the manuscript; and Nancy Nicol, for reading and help in typing the manuscript.

REFERENCES

ADAMSON, E.D. & GARDNER, R.L. (1979) Control of early development. Brit. Med. Bull. 35, 113-119.

ASHMAN, R.F., KANNO, Y. & LOWENSTEIN, W.R. (1964) The formation of a high resistance barrier in a dividing cell. Science 145, 604-605.

BENNETT, M.V.L. (1973) Function of electrotonic junctions in embryonic and adult tissues. Fed. Proc. 32, 65-75.

BENNETT, M.V.L. (1978) Junctional permeability. In: J. Feldman, N.B. Gilula and J. Pitts (Eds.), Intercellular Junctions and Synapses, Chapman and Hall, London, pp. 23-36.

BENNETT, M.V.L. & GOODENOUGH, D.A. (1978) Gap junctions, electrotonic coupling and intercellular communication. Neurosciences Research Program Bulletin, Vol. 16, M.I.T. Press, Cambridge.

BENNETT, M.V.L. & TRINKAUS, J.P. (1970) Electrical coupling between embryonic cells by way of extracellular space and specialized junctions. J. Cell Biol. 44, 592-610.

BENNETT, M.V.L., SPIRA, M.E. & PAPPAS, G.D. (1972) Properties of electrotonic junctions between embryonic cells of Fundulus. Develop. Biol. 29, 409-435.

BENNETT, M.V.L., SPIRA, M.E. & SPRAY, D.C. (1978) Permeability of gap junctions between embryonic cells of Fundulus: a reevaluation. Develop Biol. 65, 114-125.

BENNETT, M.V.L., BROWN, J.E., HARRIS, A.L. & SPRAY, D.C. (1979) Electrotonic junctions between Fundulus blastomeres: reversible block by low intracellular pH. Biol. Bull. 155, 428-429.

BLACKSHAW, S.E. & WARNER, A.E. (1976) Low resistance junctions between mesoderm cells during development of trunk muscles. J. Physiol. 255, 209-230.

BLUEMINK, J.G., VAN MAURIK, P. & LAWSON, K.A. (1975) Intimate cell contacts at the epithelial/mesenchymal interface in embryonic mouse lung. J. Ultrastruct. Res. 55, 252-270.

CEREIJIDO, M., ROBBINS, E.S., DOLAN, W.J., ROTUNNO, C.A. & SABATINI, D.D. (1978) Polarized monolayers formed by epithelial cells on a permeable and translucent support. J. Cell Biol. 77, 853-880.

COOKE, J. (1975) The emergence and regulation of spatial organization in early animal development. Ann. Rev. Biophys. Bioeng. 4, 185-217.

COOKE, J. & ZIEMAN, E.C. (1976) A clock and wavefront model for the control of the number of repeated structures during animal morphogenesis. J. Theor. Biol. 58, 455-476.

CRICK, F.H.C. (1970) Diffusion in embryogenesis. Nature 225, 420-422.

CRICK, F.H.C. & LAWRENCE, P.A. (1975) Compartments and polyclones in insect development. Science 189, 340-347.

DE HAAN, R.L. & SACHS, H.G. (1972) Cell coupling in developing systems. Curr. Top. Develop. Biol. 7, 193-228.

DELEZE, J. & LOEWENSTEIN, W.R. (1976) Permeability of a cell junction during intracellular injection of divalent cations. J. Membr. Biol. 28, 71-86.

DICAPRIO, R.A., FRENCH, A.S. & SAUNDERS, F.J. (1975) Intercellular connectivity in 8-cell Xenopus embryo. Correlation of electrical and morphological investigations. Biophys. J. 15, 373.

DIXON, J.S. & CROMLY-DILLON, J.R. (1972) The fine structure of the developing retina in Xenopus laevis. J. Embryol. Exp. Morph. 28, 659-666.

DUCIBELLA, J. & ANDERSON, E. (1975) Cell shape and membrane changes in the eight cell mouse embryo: pre-requisites for morphogenesis of the blastocyst. Develop. Biol. 47, 45-58.

EPSTEIN, M. & GILULA, N.B. (1977) A study of communication specificity between cells in culture. J. Cell Biol. 75, 769-787.

FLAGG-NEWTON, J., SIMPSON, I. & LOEWENSTEIN, W. (1979) Permeability of the cell-to-cell membrane channels in mammalian cell junctions. Science 205, 404-407.

FENTIMAN, I.S. & TAYLOR-PAPADIMITRIOU, J. (1977) Cultured human breast cells lose selectivity in direct intercellular communication. Nature 269, 151-157.

FURSHPAN, E.J. & POTTER, D.D. (1959) Transmission at the giant motor synapses of the crayfish. J. Physiol. 145, 289-325.

FURSHPAN, E.J. & POTTER, D.D. (1968) Low-resistance junctions between cells in embryos and tissue culture. Curr. Top. Develop. Biol. 3, 95-127.

GARCIA-BELLIDO, A. (1975) Genetic control of wing disc development in Drosophila. In: Cell Patterning, Ciba Foundation Symposium, 29, 161-182.

GARDNER, R.L. & ROSSANT, J. (1976) Determination during embryogenesis. In: Embryogenesis in Mammals, Ciba Foundation Symposium, 40, 5-25.

GAUNT, S.J. & PAPAIOANNOU, V.E. (1979) Metabolic cooperation between embryonic and embryonal carcinoma cells of the mouse. J. Embryol. Exp. Morph. 54, 263-275.

GIERER, A. & MEINHARDT, H. (1972) A theory of biological pattern formation. Kybernetik 12, 30-39.

GILULA, N.B. (1974) Junctions between cells. In: R.P. Cox (Ed.), Cell Communication, John Wiley & Sons, New York, pp. 1-29.

GILULA, N.B. (1977) Gap junctions and cell communication. In: B.R. Brinkley and K.R. Porter (Eds.), International Cell Biology, Rockefeller University Press, New York, pp. 61-69.

GILULA, N.B., REEVES, R.O. & STEINBACK, A. (1972) Metabolic coupling, ionic coupling and cell contact. Nature 235, 262-265.

GILULA, N.B., EPSTEIN, M.L. & BEERS, W.H. (1978) Cell-to-cell communication and ovulation. A study of the cumulus-oocyte complex. J. Cell Biol. 78, 58-75.

GOODWIN, B.C. & COHEN, M.H. (1969) A phase shift model for the spatial and temporal organization of developing systems. J. Theor. Biol. 25, 49-107.

HANNA, R.B., SPRAY, D.C., MODEL, P.G., HARRIS, A.L. & BENNETT, M.V.L. (1979) Ultrastructural and physiology of gap junctions of an amphibian embryo. Effects of CO_2. Biol. Bull. 155:442.

HOLTFRETER, J. & HAMBURGER, V. (1955) In: Wieler, Weiss and Hamburger (Eds.) Analysis of Development, Saunders, Philadelphia, pp. 230-296.

HORSTADIUS, S. (1973) Experimental embryology of echinoderms. Clarendon Press, Oxford.

HSU, YU-CHIH (1979) In vitro development of individually cultured whole mouse from blastocyst to early somite stage. Develop. Biol. 68, 453-461.

ITO, S. & HORI, N. (1966) Electrical characteristics of Triturus egg cells during cleavage. J. Gen. Physiol. 49, 1019-1027.

ITO, S. & LOEWENSTEIN, W.R. (1969) Ionic communication between early embryonic cells. Develop. Biol. 19, 228-243.

JOHNSON, R.G. & SHERIDAN, J.D. (1971) Junctions between cancer cells in culture: ultrastructure and permeability. Science 174, 717-719.

JOHNSON, M.H., CHAKRABORTY, J., HANDYSIDE, A.H., WILLISON, K., & STERN, P. (1979) The effect of prolonged decompaction on the development of the pre-implantation mouse embryo. J. Embryol. Exp. Morph. 54, 263-275.

KANNO, Y. & LOEWENSTEIN, W.R. (1966) Cell-to-cell passage of large molecules. Nature, 212, 629-630.

KEMLER, R., BABINET, C., EISEN, H. & JACOB, F. (1977) Surface antigen in early differentiation. Proc. Natl. Acad. Sci. 74, 449-4452.

LAWRENCE, P.A. (1973) A clonal analysis of segment development in Oncopeltus (Hemiptera). J. Embryol. Exp. Morph. 30, 681-699.

LAWRENCE, P.A. & GREEN, S.M. (1975) The anatomy of a compartment border. The intersegmental boundary in Oncopeltus. J. Cell Biol. 65, 373-382.

LAWRENCE, P.A., CRICK, F.H.C. & MUNRO, M. (1972) A gradient of positional information in an insect, Rhodnius. J. Cell Sci. 11, 815-853.

LAWRENCE, T.S., BEERS, W.H. & GILULA, N.B. (1978) Transmission of hormonal stimulation by cell-to-cell communication. Nature 272, 501-506.

LEHTONEN, E. (1975) Epithelio-mesenchymal interface during mouse kidney tubule induction in vivo. J. Embryol. Exp. Morph. 34, 695-705.

LEHTONEN, E. (1976) Transmission of signals in embryonic induction. Med. Biol. 54, 108-128.

LO, C.W. & GILULA, N.B. (1979a) Gap junctional communication in the pre-implantation mouse embryo. Cell 18, 399-409.

LO, C.W. & GILULA, N.B. (1980a) PCC4azal teratocarconoma stem cell differentiation in culture. II. Morphological characterization. Develop. Biol. 74, in press.

LO, C.W. & GILULA, N.B. (1980b) PCC4azal teratocarcinoma stem cell differen-tiation in culture. III. Cell-to-cell communication properties. Develop. Biol. 74, in press.

LOCKE, M. (1959) The cuticular pattern in an insect, Rhodnius prolixus. J. Exp. Biol. 36, 459-477.

LOCKE, M. (1960) The cuticular pattern in an insect. The intersegmental membranes. J. Exp. Biol. 37, 398-406.

LOCKE, M. (1967) The development of patterns in the integument of insects. Adv. Morphogen. 6, 33-88.

LOEWENSTEIN, W.R. (1966) Permeability of membrane junctions. Ann. N.Y. Acad. Sci. 137, 441-472.

LOEWENSTEIN, W.R. (1968) Communication through cell junctions: Implications in growth control and differentiation. Develop. Biol. (Suppl. 2), 151-183.

LOEWENSTEIN, W.R. (1975) Permeable junctions. Cold Spring Harbor Symp. Quant. Biol. 40, 49-63.

LOEWENSTEIN, W.R. (1979) Junctional intercellular communication and the control of growth. Biochim. Biophys. Acta 560, 1-65.

LOPRESTI, V., MACAGNO, E.R. & LEVINTHAL, C. (1974) Structure and development of neuronal connections in isogenic organisms: Transient gap junctions between growing optic axons and lamina neuroblast. Proc. Natl. Acad. Sci. U.S.A. 71, 1089-1102.

MAGNUSON, T., ANTHONY, D. & STACKPOLE, C.W. (1977) Characterization of intercellular junctions in the preimplantation mouse embryo by freeze fracture and thin section electron microscopy. Develop. Biol. 61, 252-261.

MARTIN, G.R. (1975) Teratocarcinomas as a model system for the study of embryogenesis and neoplasia. Cell 5, 229-243.

78

McMAHON, D. & WEST, C. (1976) Transduction of positional information during development. In: C.G. Poste and G.L. Nicolson (Eds.), The Cell Surface in Animal Embryogenesis and Development, Elsevier/North Holland, Amsterdam, pp. 449-482.

McNUTT, N.S. & WEINSTEIN, R.S. (1970) The ultrastructure of the nexus. A correlated thin section and freeze-cleave study. J. Cell Biol. 47, 666-688.

McNUTT, N.S. & WEINSTEIN, R.S. (1973) Membrane ultrastructure at mammalian intercellular junctions. Prog. Biophys. Mol. Biol. 26, 45.

MICHALKE, W. (1977) A gradient of diffusible substance in a monolayer of cultured cells. J. Membr. Biol. 33, 1-20.

MICHALKE, W. & LOEWENSTEIN, W.R. (1971) Communication between cells of different types. Nature 232, 121-122.

MINTZ, B. (1967) Gene control of mammalian pigmentary differentiation. I. Clonal origin of melanocytes. Proc. Natl. Acad. Sci. U.S.A. 58, 344-351.

MINTZ, B. (1971) Clonal basis of mammalian differentiation. Symp. Soc. Exp. Biol. 25, 345-369.

MINTZ, B. & ILLMENSEE, U. (1975) Normal genetically mosaic mice produced from malignant teratocarcinoma cells. Proc. Natl. Acad. Sci. U.S.A. 72, 3585-3589.

MUNRO, M. & CRICK, F.H.C. (1971) The time needed to set up a gradient: Detailed calculations. Symp. Soc. Exp. Biol. 25, 439-454.

OLIVEIRA-CASTRO, G.M. & LOEWENSTEIN, W.R. (1971) Junctional membrane permeability. Effects of divalent cations. J. Membr. Biol. 5, 51-77.

PAYTON, B.W., BENNETT, M.V.L. & PAPPA, G.D. (1969) Permeability and structure of junctional membrane at an electrotonic synapse. Science 166, 1641-1643.

PEDERSEN, R.A. & SPINDLE, A.I. (1980) The role of the blastocoele micro-environment in early mouse embryo differentiation. Nature, in press.

PITTS, J.D. (1972) Direct interaction between animal cells. In: L.G. Silvestri (Ed.), Cell Interactions, Elsevier/North Holland, Amsterdam, pp. 277-285.

PITTS, J.D. (1977) Direct communication between animal cells. In: B.R. Brinkley and K.R. Porter (Eds.), International Cell Biology. 1976-1977, Rockefeller University Press, New York, pp. 43-49.

PITTS, J.D. (1978) Junctional communication and cellular growth control. In: J. Feldman, N.B. Gilula and J.D. Pitts (Eds.), Intercellular Junctions and Synapses, Chapman and Hall, London, pp. 61-80.

PITTS, J.D. & BURK, R.R. (1976) Specificity of junctional communication between animal cells. Nature 264, 762-764.

PITTS, J.D. & SIMMS, J.W. (1977) Permeability of junctions between animal cell. Intercellular transfer of nucleotides but not of macromolecules. Exp. Cell Res. 104, 153-163.

POTTER, D.D., FURSHPAN, E.J. & LENNOX, E.S. (1966) Connections between cells of the developing squid as revealed by electrophysiological methods. Proc. Natl. Acad. Sci. U.S.A. 55, 328-336.

REESE, T.S., BENNETT, M.V.L. & FEDER, N. (1971) Cell-to-cell movement of peroxidases injected into the septate axon of crayfish. Anat. Rec. 169, 409.

79

ROBINSON, K.R. & JAFFE, L.F. (1973) Ion movements in a developing fucoid egg. Develop. Biol. 35, 349-361.

ROSE, B. & LOEWENSTEIN, W.R. (1975) Permeability of cell junctions depends on local cytoplasmic calcium activity. Nature 254, 250-252.

ROSE, B. & LOEWENSTEIN, W.R. (1976) Permeability of a cell junction and the local cytoplasmic free ionized calcium concentration: A study with aequorin. J. Membr. Biol. 28, 87-119.

ROSE, B., SIMPSON, I. & LOEWENSTEIN, W.R. (1977) Calcium ion produces graded changes in permeability of membrane channels in cell junctions. Nature 267, 625-627.

ROSSANT, J. (1978) Cell commitment in early rodent development. In: M.H. Johnson (Ed.), Development in Mammals, North Holland, Amsterdam, pp. 119-150.

SATIR, P. & GILULA, N.B. (1973) The fine structure of membranes and inter-cellular communication in insects. Ann. Rev. Entomol. 18, 143-166.

SAXEN, L., LEHTONEN, E., KARKINEN-JAASKELAINEN, M., NORDLING, S. & WARTIOVAARA, WARTIOVAARA, J. (1976) Are morphogenetic tissue interactions mediated by transmissible signal substances or through cell contacts? Nature 259, 662-663.

SHERIDAN, J.D. (1968) Electrophysiological evidence for low-resistance intercellular junctions in the early chick embryo. J. Cell Biol. 37, 650-659.

SHERIDAN, J.D. (1971) Dye movement and low resistance junctions between reaggregated embryonic cells. Develop. Biol. 26, 627-636.

SHERIDAN, J.D. (1974) Low resistance junctions: Some functional considerations siderations. In: A.A. Moscona (Ed.), The Cell Surface in Development, Wiley, New York, pp. 187-206.

SHERIDAN, J.D. (1976) Cell coupling and cell communication during embryo-genesis. In: G. Poste and G.L. Nicolson (Eds.), The Cell Surface in Animal Embryogenesis, Elsevier, Amsterdam, pp. 409-447.

SHERIDAN, J.D. (1978) Function, formation and experimental modification. In: J. Feldman, N.B. Gilula and J. Pitts (Eds.), Intercellular Junctions and Synapses, Chapman and Hall, London, pp. 37-60.

SIMPSON, I., ROSE, B. & LOEWENSTEIN, W.R. (1977) Size limit of molecules permeating the junctional membrane channels. Science 195, 294-296.

SLACK, C. & PALMER, J.F. (1969) The permeability of intercellular junctions in the early embryo of Xenopus laevis studied with a fluorescent tracer. Exp. Cell Res. 55, 416-419.

SLAVKIN, H.C. & BRINGAS, P. Jr. (1976) Epithelial-mesenchyme interactions during odontogenesis. IV. Morphological evidence for direct heterotypic cell-cell contacts. Develop. Biol. 50, 428-442.

SPEMANN, H. (1938) Embryonic Development and Induction, Yale University Press, New Haven.

SPRAY, D.C., HARRIS, A.L. & BENNETT, M.V.L. (1979) Voltage dependence of junctional conductance in early amphibian embryos. Science 204, 432-434.

STAEHELIN, L.A. (1974) Structure and function of intercellular junctions. Int. Rev. Cytol. 39, 191-278.

STEVENS, L.C. (1967) The biology of teratomas. Adv. Morphogen. 6, 1-32.

STOKER, M. (1975) The effects of topoinhibition and cytochalasin B on metabolic cooperation. Cell 6, 253-257.

SUBAK-SHARPE, H., BURK, R.R. & PITTS, J.D. (1966) Metabolic cooperation by cell-to-cell transfer between genetically different mammalian cells in tissue culture. Heredity 21, 342-343.

SUMMERBELL, D., LEWIS, J.H. & WOLPERT, L. (1973) Positional information in chick limb morphogenesis. Nature 244, 492-496.

TACHI, S., TACHI, C. & LINDER, H.R. (1970) Ultrastructural features of blastocyst attachment and trophoblastic invasion in the rat. J. Reprod. Fertil. 21, 37-56.

TARKOWSKI, A.K. & WROBLEWSKA, J. (1967) Development of blastomeres of mouse eggs isolated at the four- and eight-cell stage. J. Embryol. Exp. Morphol. 18, 155-180.

TUPPER, J.T. & SAUNDERS, J.W. Jr. (1972) Intercellular permeability in the early Asterias embryo. Develop. Biol. 27, 546-554.

TUPPER, J.T., SAUNDERS, J.W. & EDWARDS, C. (1970) The onset of electrical communication between cells in the developing starfish embryo. J. Cell Biol. 46, 187.

TURIN, L. & WARNER, A. (1977) Carbon dioxide reversibly abolishes ionic communication between cells of early amphibian embryo. Nature 270, 56-57.

WARNER, A. (1973) The electrical properties of the ectoderm in the amphibian embryo during induction and early development of the nervous system. J. Physiol. 235, 267-286.

WARNER, A.E. & LAWRENCE, P.A. (1973) Electrical coupling across developmental boundaries in insect epidermis. Nature 245, 47-48.

WARTIOVAARA, J., NORDLING, S., LEHTONEN, E. & SAXEN, L. (1974) Transfilter induction of kidney tubules: Correlation with cytoplasmic penetration into Nucleopore filters. J. Embryol. Exp. Morphol. 31, 667-682.

WOLPERT, L. (1969) Positional information and the spatial pattern of cellular differentiation. J. Theor. Biol. 25, 1-47.

WOLPERT, L. (1971) Positional information and pattern formation. Curr. Top. Develop. Biol. 6, 183-224.

WOLPERT, L. (1978) Gap junctions: Channels for communication in development. In: J. Feldman, N.B. Gilula and J. Pitts (Eds.), Intercellular Junctions and Synapses, Chapman and Hall, London, pp. 83-94.

WOODWARD, D.J. (1968) Electrical signs of new membrane production during cleavage of Rana pipiens eggs. J. Gen. Physiol. 52, 509-531.

© 1980 Elsevier/North-Holland Biomedical Press
Development in Mammals, Vol. 4, M.H. Johnson, editor

PLASMINOGEN ACTIVATOR IN EARLY DEVELOPMENT

Sidney Strickland

The Rockefeller University
New York, New York 10021
U.S.A.

1.   General properties of plasminogen activators

2.   Ovulation

3.   Spermatogenesis

4.   Implantation

5.   Early embryogenesis

6.   Conclusions

'The detail of the pattern is movement'

T.S. Eliot, "Burnt Norton"

A student of morphogenesis may be compared to a frustrated
movie fan reduced to looking at films one frame at a time –
movement can be sensed and charted, but a feeling for the dynamic
nature of the event is lost. Thus, it is easy to lose sight of
the fact that in embryogenesis, degradative as well as synthetic
processes abound. The embryo, with no respect for landmark status,
builds elaborate structures only to tear them down; with no
feeling for territorial rights, columns of cells on their way to
precisely defined destinations push through everything on their
path. As the world's most extraordinary and complex architecture
is assembled, there is in fact wholesale destruction.

In the course of tissue remodeling, many of the molecules
that must be destroyed first are extracellular proteins. In
histological examinations of these processes, it is often found
that a basement membrane (primarily type IV collagen and non-
collagen glycoproteins) first shows deterioration, followed in
many instances by death of the attached cells.

Certain predictions can thus be formulated about the mechanism by which cells invade and destroy tissues: 1) it is likely that proteolytic enzymes are involved; 2) the pH optima of these proteases should be in the neutral range observed in the extracellular fluid; 3) the production of these enzymes should be precisely timed; and 4) there should exist a means to restrict the proteolytic action in space and time.

The first detailed description of a link between tissue remodeling and proteases came from Gross (1974). He found that the resorption of the tadpole tail during metamorphosis was correlated with the copious production of a collagenolytic activity (Gross & Lapiere, 1962). Only certain tissues had this activity, namely those which consist of relatively large amounts of connective tissue and which undergo dramatic structural alterations during metamorphosis (Usuku & Gross, 1965). However, since animal collagenases do not cleave non-collagen proteins, and only make a single cleavage through triple-helical collagen (Gross & Nagai, 1965; Sakai & Gross, 1967), it is apparent that more than this one enzyme is required for extensive degradation of the extracellular matrix.

In this review I will summarize the evidence that the protease plasminogen activator participates in several examples of tissue degradation in reproduction and development. I will also focus on the usefulness of this enzyme as a marker for studies on the hormonal control of developmental processes.

I would like first to register several disclaimers. This contribution will not comprehensively cover the extensive literature on plasminogen activator, but is restricted to developmental systems (I hope purists will forgive, in an era of specialization, the inclusion of ovulation and spermatogenesis in a contribution on development). In addition, I will not discuss every protease that has been documented in embryogenesis, but concentrate on the relatively recent work on plasminogen activator.

1.   GENERAL PROPERTIES OF PLASMINOGEN ACTIVATORS

Plasminogen activators are serine proteases capable of converting plasminogen to plasmin. Whereas plasminogen has no catalytic activity, plasmin is itself a serine protease of broad specificity. The activation of plasminogen occurs via cleavage of an arginine-valine bond, which converts the single chain molecule into the two chain plasmin (see Christman *et al*, 1977, for review).

Operationally, plasminogen activators are defined as follows: at certain concentrations of these enzymes, no proteolysis can be observed, whereas addition of plasminogen to the assay results in extensive proteolysis. It should be noted that these conditions are not met by many serine proteases with arginine/lysine specificity. For example, although trypsin is capable of cleaving the arginine-valine bond in plasminogen mentioned above, it also extensively degrades the molecule by hydrolysing other peptide bonds. Consequently, the stimulation of tryptic proteolytic activity by plasminogen addition is only several-fold, in contrast to the essentially "all or none" effect with plasminogen activators.

It is important to emphasize the attractiveness of the plasminogen activator/plasminogen system as a means for generalized extracellular proteolysis. Some of the most obvious advantages are as follows:

1. Both plasminogen activator and plasminogen have a pH optimum near neutrality. This property ensures maximum activity at the pH of the extracellular fluid. In contrast, many of the lysosomal proteases are poor candidates for extracellular action, since they have very low activity in the neutral range.

2. The catalytic amplification of proteolysis inherent in the activation of plasminogen is crucial. In many developmental systems, relatively few cells must have an impact on large amounts of tissue. For a cell to make enough enzyme to fulfill this purpose would be difficult. This concept of proteolytic cascades also governs the blood coagulation and complement pathways.

3. The amount of plasminogen in the plasma of birds and mammals is approximately 0.5 milligrams per millilitre (Schultze & Heremans, 1966). This amount represents an enormous and almost inexhaustible supply of proenzyme

for processes occurring in various parts of the
organism.

4. Plasmin is a protease that can extensively degrade an
    extremely broad range of proteins, being exceeded in
    generality of action only by the digestive enzymes. In
    contrast, serine proteases of the complement and
    coagulation systems have restricted substrate speci-
    ficity. Thus, plasmin is well-designed to attack a
    variety of substrates. Native collagen, which represents
    a substantial proportion of the extracellular matrix, is
    one molecule that is not susceptible to the action of
    plasmin. However, after the action of animal collagenase
    on collagen, the reaction products (two helical fragments)
    denature about 5 degrees below body temperature, becoming
    a substrate for plasmin (Sakai & Gross, 1967). One can
    therefore visualize various ways in which plasmin might
    be involved in degradation of collagen. One is that
    catalytic amounts of collagenase might be produced with
    plasminogen activator, enabling plasmin to hydrolyze
    both collagen and non-collagen proteins. A second possi-
    bility is that a zymogen form of collagenase (Harris &
    Krane, 1974) is present or secreted, and that this
    proenzyme can be catalytically activated by plasmin. In
    this case, plasminogen activator would generate proteo-
    lytic activity (plasmin), which would in turn generate
    collagenolytic activity. This latter possibility is
    strengthened by the observation that rheumatoid synovial
    cells produce both a procollagenase and plasminogen
    activator, and that plasmin is capable of activating the
    procollagenase (Werb *et al*, 1977).

5. The generation of plasmin would obviously be limited to
    those locations where plasminogen activators and
    plasminogen co-existed. After generation, the action of
    plasmin would be severely restricted in space and time
    by the plasma protease inhibitors, primarily alpha-2
    macroglobulin and alpha-1 antitrypsin. The precision of
    degradative events is a hallmark of embryogenesis, and
    is obviously of fundamental importance. To achieve such

precision, rapid attenuation of generated proteolytic
activity is imperative.

One classic example of controlled tissue degradation that has
been shown to involve proteases is the hatching of the silk moth
(Kafatos, 1972). Although not a mammalian system, it is worth-
while to consider briefly the mechanism by which the moth escapes
its cocoon. The cocoon is dissolved at hatching by the protease
cocoonase. This enzyme is secreted as a proenzyme, and subsequently activated by proteolytic cleavage. The catalytic specificities and inhibitor profiles of these proteins both suggest
that the activator is analogous to plasminogen activator, and
that cocoonase is related to plasmin. One of the interesting
features about this system is that proteolysis is determined by
the delivery of the proenzyme, rather than the activator. It is
important to keep this fact in mind when considering mechanisms
for tissue remodeling in mammals - extensive proteolysis requires
the simultaneous presence of plasminogen and plasminogen activator,
and can be controlled by the availability of either component.

In summary, by secreting a small amount of plasminogen
activator, a cell can generate considerable proteolytic activity
of broad specificity that is precisely localized. There are
other features that suggest the generality of this system, for
example, the hormonal control of secretion. The remainder of this
article discusses several processes in which the secretion of
plasminogen activator has been shown to be correlated with an
event that requires the degradation of extracellular protein.

2.   OVULATION

The ovarian follicle at the time of ovulation consists of a
fluid filled antrum surrounded by a layer of granulosa cells; the
ovum rests on a stalk of granulosa cells called the cumulus
oophorus. The antrum is filled with a filtrate of blood called
follicular fluid. No blood vessels penetrate into the granulosa
cell layer, so that these cells and the ovum reside in an
avascular space. Surrounding the granulosa cell layer is a basement membrane, and around this several layers of thecal cells
(theca interna and theca externa). A schematic representation of
a Graafian follicle is shown in Figure 1.

At the time of ovulation, the following events take place:
the ovum and its investments are dislodged from the granulosa
cell layer and become free-floating in the follicular fluid
(Brambell, 1956). The granulosa cell layer is disrupted, and in
the region of the wall that is destined to rupture, the entire
layer is removed (Parr, 1974). Finally, the basement membrane
and thecal layers, which comprise the remainder of the follicle
wall at the point of rupture, are degraded. The wall becomes
thin, a stigma forms, ultimately tears, and the follicular contents including the ovum ooze out.

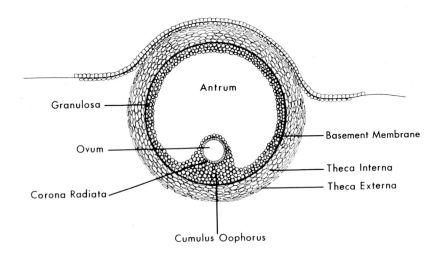

Figure 1: Schematic representation of an ovarian Graafian follicle.

Ovulation ranks as one of the earliest biological processes
systematically examined. Over 300 years ago, Regnier de Graaf
(1672) provided a detailed and accurate description of the ovarian
follicle and the events that accompany its rupture. Many
hypotheses have been put forward since that time to explain this
phenomenon. They include the following:

1.   Intrafollicular pressure increases until it exceeds
     the strength of the follicle wall, and the follicle
     explodes.

2.   The swelling of the follicle restricts the blood
     supply to the wall, which then degenerates and tears.

These theories have been reviewed, and the arguments against them
catalogued by Espey (1974).

An enzymatic theory of ovulation was first proposed by
Schochet in 1916. His paper is astonishing for its prescience.
For example, his study was undertaken to determine "whether the
liquor folliculi has a digestive action and if so does it
possess a specific enzyme that can be demonstrated by dialysis or
other tests, [and] if it possesses such action, under what conditions is it altered?"

Schochet's protocol involved taking various porcine fluids,
including follicular and amniotic, and incubating them in a
dialysis bag with proteinaceous subtrates, e.g. fibrin. Proteolytic activity was measured by the amount of ninhydrin-positive
material in the dialysate. It is interesting to note that the
method of fibrin preparation used by him would have resulted in a
product heavily contaminated with plasminogen. His results
showed that using fibrin as a substrate, follicular fluid from
Graafian follicles contained high levels of proteolytic activity,
in contrast to amniotic fluid which had none. Schochet's conclusion was that "the rupture of the Graafian follicle is due in
part to the digestion of the theca folliculi by a proteolytic
ferment or enzyme in the liquor folliculi".

It is a paradigm of modern biochemistry that to do clean
experiments, one must use clean reagents. The failure of the
attempts to reproduce Schochet's results illustrate one example in
which purity of reagents was disadvantageous. Schochet was
obviously measuring plasminogen activator and/or plasmin (see
below), and was unwittingly aided by the use of a substrate loaded
with plasminogen. His successors used more purified protease
substrates; the absence of plasminogen in their assays precluded
the amplification mentioned above, and made demonstration of
proteolysis difficult.

The next demonstration of ovarian proteolytic activity was by
Reichert (1962). He observed two activities, one with pH optimum
around 8 and the other one around 4. The "pH 8 protease" was
increased by treatment of the rats with follicle stimulating

hormone (FSH). In the light of the pH profile of this enzyme, its inhibitor profile, and its induction by FSH, it is likely that Reichert was measuring plasmin (Beers, 1975; see below).

The banner of a proteolytic mechanism of ovulation was then taken up by Espey. His experiments showed that several serine proteases, e.g. trypsin, could weaken strips of follicle wall *in vitro* (Espey, 1970). Furthermore, injection of trypsin into Graafian follicles resulted in rupture in a manner that mimicked ovulation (Espey & Lipner, 1965).

Although the granulosa cells exist in an avascular compartment, the follicular fluid is composed primarily of molecules derived from blood (Caravaglios & Cilotti, 1956; Perloff *et al*, 1955). The barrier that separates granulosa from thecal cells acts as a molecular sieve, with proteins of around 250,000 daltons present in follicular fluid at 50% of their plasma concentrations (Shalgi *et al*, 1973). This fact indicates that various proteins that can influence proteolytic activity, e.g. certain protease inhibitors and plasminogen, should be present in follicular fluid at approximately plasma concentrations. This expectation has been confirmed experimentally by Beers (1975). In addition, he demonstrated that the protease plasmin is present in follicular fluid and in follicle wall homogenates, and that plasmin, like trypsin, can weaken follicle wall strips *in vitro*.

These various experimental threads have been drawn together by the realization that under certain conditions granulosa cells synthesize and secrete plasminogen activator. The pattern of enzyme formation in the ovary is consistent with a fundamental role for plasminogen activator in follicle rupture (Beers, Strickland & Reich, 1975; Strickland & Beers, 1976). Briefly, the results can be summarized as follows:

1. As ovulation approaches, the granulosa cells of the rat follicle produce increasing amounts of plasminogen activator. During the eight hours preceding ovulation, the increase in enzyme synthesis and secretion is more than 100-fold with the peak of activity coinciding with follicle rupture. After this time, there is a rapid decrease in enzyme production (Beers & Strickland, 1978a).

89

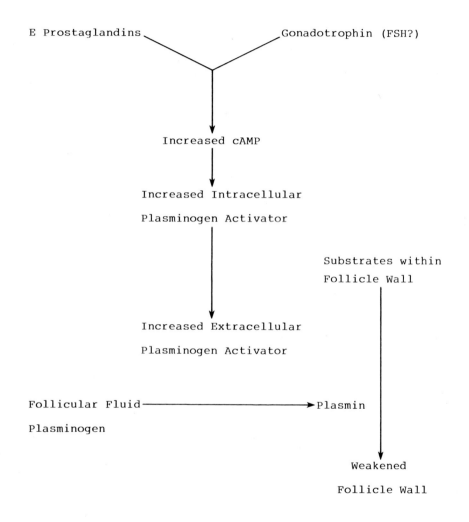

Figure 2: Hypothesis for the mechanism and
          hormonal control of ovulation.

2. When cells are collected separately from different follicles in the same ovary, those from pre-ovulatory follicles secrete enzyme whereas those from well-developed, non-preovulatory follicles do not.

3. By assaying single cells for plasminogen activator and then examining the active cells by electron microscopy, the active cell type has unambiguously been shown to be granulosa.

4. Inactive granulosa cells can be stimulated to produce plasminogen activator *in vitro* by physiological concentrations of gonadotrophins. Studies on the specificity of the response indicate that FSH is approximately 10,000-fold more effective than luteinizing hormone (LH).

5. The effects of FSH can be duplicated by analogues of cAMP and by prostaglandins E1 and E2.

Based on these correlations, we have proposed the working hypothesis summarised in Figure 2 for the role of plasminogen activator in ovulation (Strickland & Beers, 1979). Plasmin, in addition to a direct action on the follicle wall, could also activate other zymogens, e.g. procollagenase, that would contribute to eventual follicle rupture.

Recently, Pesek and Beers (1980) have shown that in contrast to its ineffectiveness in stimulating plasminogen activator production in granulosa cell cultures, LH is active in inducing granulosa cells in organ cultures of rat ovary. The action of LH is blocked by indomethacin, suggesting that the gonadotrophin may be acting to stimulate synthesis of prostaglandins, which in turn act on the granulosa cells.

The hormonal control of plasminogen activator synthesis in granulosa cells is worth considering in general terms. Given the proposition that this enzyme is of fundamental importance for ovulation, the hormonal control of protease secretion can be used as a paradigm to investigate the control of follicle rupture. Thus, our experiments have suggested a more direct role for FSH in ovulation than previously recognized. This approach can be extended to other physiological processes, e.g. embryo

implantation, for which the endocrine control is controversial.
In addition, the modulation of enzyme synthesis by hormones (or
other effectors) can provide an exquisite assay for the molecules
in question. This assay is biological yet can be performed on
amounts of material equivalent to those required for radio-
immunoassay. For FSH, it has led to the development of a cell
culture assay that is approximately a million-fold more sensitive
than the previous bioassay (Steelman & Pohley, 1953), and which
can detect as little as 15 picograms of FSH (Beers & Strickland,
1978b). For structure-activity studies, purifications, or
investigations into the mechanism of hormone action, such an
assay should prove useful.

3. SPERMATOGENESIS

Gametogenesis is analogous in some respects in the male and
female. As the ovum must escape the follicle, so the premeiotic
germ cells in the testis must transverse the tight junctions of
adjacent Sertoli cells in order to gain entry into the adluminal
compartment of the seminiferous tubule. It is not clear by what
mechanism the germ cells accomplish this end. Using the concepts
described in the preceding section as a guide, one might postulate
that helper cells analogous to granulosa in the ovary might provide proteolytic enzymes to facilitate germ cell movement. The
best candidates would be the testicular counterparts of granulosa
cells, the Sertoli cells, which nurse and maintain the development
of the germ cells.

Lacroix, Smith and Fritz (1977) have demonstrated that
cultures enriched for Sertoli cells do produce plasminogen
activator. In addition, the secretion of enzyme is stimulated by
FSH and dibutyryl cAMP. They have postulated that the enzyme may
facilitate movement of germ cells through the junctions, or
perhaps assist in the release of mature spermatids. Thus, the
functional analogies between granulosa and Sertoli cells are
mirrored in biochemical and endocrine analogies.

4. IMPLANTATION

The first differentiated cell to arise in mammalian embryogenesis is the trophectoderm, which is derived from the outer cell

layer of the preimplantation embryo. It is well established
that it is derivatives of the trophectoderm cells, the trophoblast, that are responsible for attachment of the embryo to the
uterine wall during embryo implantation. This process in many
mammals is an invasive event, in which the embryo penetrates
through the uterine epithelium and its underlying basement
membrane and finally breaches maternal blood vessels before
advancement ceases. Pathologists have often drawn attention to
the degradative nature of these cells by pointing out their
"pseudomalignant" properties (Novak & Woodruff, 1974). In fact,
the normal trophoblast is capable of physiologic metastasis,
since these cells can be found in the lungs of nearly half of
all pregnant women (Novak & Woodruff, 1974).

In a great majority of cases, the trophoblast does no
permanent damage to the maternal host. One possible reason for
this is that the uterus responds to the invading cells with a
local defence mechanism called the decidual reaction, in which the
uterine wall becomes impenetrable to the trophoblast. However,
this is not the only factor responsible for limiting invasion,
since it is known that trophoblast transferred to ectopic sites
such as the testis are invasive only during a limited portion
of their life span (Kirby, 1965). Consequently, the destructive
properties of the normal trophoblast appear to be controlled
carefully. Since these cells are reported to be among the most
invasive of all cells known, and can penetrate and destroy neoplastic as well as normal tissue (Solomon, 1966), it is not
surprising that pathology of the trophoblast has deadly consequences. Choriocarcinoma, in which the trophoblast remains
invasive throughout and beyond pregnancy, is one of the most
metastatic tumours described, and before the advent of chemotherapy
was virtually always fatal (Romney et al, 1975).

In view of these interesting characteristics, it is perhaps
surprising that little is known about the biochemical basis for
trophoblast invasion. Various investigators have reported the
association of proteolytic enzymes with implanting embryos
(Owers & Blandau, 1971; Denker, 1972; Kirchner, 1972; Pinsker et
al, 1974), but none of these studies characterized the enzyme or
demonstrated a correlation with the process of implantation.

Since the tissue destruction that accompanies implantation can
be regarded as analogous to that associated with ovulation, and
since as noted above, plasminogen activator is involved in
follicular rupture, we investigated embryos at the time of
implantation for the production of this protease (Strickland *et
al*, 1976). Our results can be summarized as follows:

1. Blastocysts collected from mice on the fourth day of
   pregnancy were cultured *in vitro* for four days. The
   resultant blastocyst outgrowth produced fibrinolytic
   zones when assayed by the fibrin-agar overlay technique.

2. The fibrinolytic activity was strictly plasminogen
   dependent, and hence due to plasminogen activator in the
   conditioned medium. Thus, the activity was abolished
   when the assays were performed in the absence of
   plasminogen, and full activity was restored by addition
   of purified plasminogen.

3. The level of activity was proportional to the number of
   blastocysts in the culture.

4. The time course of enzyme production in cultures of second,
   third, fourth or fifth day embryos was a function of
   gestation age and not of time in culture.

5. The onset of enzyme secretion was not directly related to
   the emergence of embryos from the zona pellucida. Blastocysts (fourth day) which normally require about 30 hours
   to hatch spontaneously *in vitro*, can be artificially freed
   from their zonae pellucidae by treatment with pronase
   (Mintz, 1962). When blastocysts hatched with pronase on
   the fourth day were cultured and assayed for plasminogen
   activator, the time course of enzyme production was
   indistinguishable from that of untreated cultures.

6. Elevated levels of enzyme were present intracellularly as
   well as in the conditioned medium. The enzyme was
   undetectable in cell lysates before equivalent gestation
   day 6; subsequently, the increase in levels of intracellular plasminogen activator paralleled roughly that
   assayed in conditioned medium.

7. The temporal pattern of plasminogen activator production by the blastocyst cultures was complex. Initial enzyme activity appeared on about the sixth day, rising to a temporary maximum on the eighth or ninth day; this was followed by a transient decline with a second and continuing increase starting at the eleventh day. The second phase of enzyme secretion was maintained until at least the fifteenth day, by which time the level of activity was five-fold higher than the plateau value at the eighth day. This pattern suggested the participation of two types of cells, with the first initiating enzyme production by the sixth day and declining after reaching a maximum by the ninth day, and the second beginning to secrete detectable amounts at a later time with progressive increases thereafter.

8. The pattern of plasminogen activator production *in vitro* by blastocysts that had been delayed from implanting by maternal ovariectomy was indistinguishable from control blastocysts.

9. Fractionation of the blastocyst into its constituent cell types has identified the trophoblast as the cell responsible for the first phase of enzyme synthesis. Beginning on the seventh day, pure trophoblast cultures *in vitro* produced increasing amounts of plasminogen activator until the ninth day, after which enzyme secretion gradually decreased.

The invasive period of the mouse trophoblast has been determined primarily from transplantation of embryos to ectopic sites, and it extends from the sixth to the tenth equivalent day of gestation (Kirby, 1965). This interval corresponds closely with the period during which the cultured trophoblast is secreting plasminogen activator, and suggests that the enzyme is a contributing factor in implantation.

Comparative studies performed with the pig embryo lend support to this conclusion (Mullins *et al*, 1980). In the pig, there is no invasion of the uterine endometrium by the embryo. The type of placenta formed is epitheliochorial, in which all the maternal tissues are preserved intact, and nutrient exchange relies on interdigitation of microvilli on the opposing epithelial surfaces of the trophoblast and uterus (Amoroso, 1952). The pig trophoblast is capable of tissue invasion, however, if transferred to an ectopic site (Samuel, 1971; Samuel & Perry, 1972). These facts suggest that the uterus may produce specialized molecules or structures that restrict the ability of the pig embryo to implant. Mullins *et al* (1980) have shown that the pig blastocyst does in fact produce plasminogen activator, which may account for its invasive properties in extrauterine sites. They have further demonstrated that the uterus at the time of placentation produces an inhibitor of the enzyme; this inhibitor, which is clearly distinct in several respects from serum protease inhibitors, appears to be induced *in vivo* by progesterone.

Taken together, studies on the establishment of the maternal-embryo relationship in mouse and pig both support the concept that plasminogen activator plays a central role in the ability of the trophoblast to invade other tissues. The discovery of molecules that could modulate enzyme production by the tropho-blast would be of considerable interest, since such knowledge could shed light both on the role of plasminogen activator in implantation, and on the design of rational strategies for the prevention of embryo attachment.

5. EARLY EMBRYOGENESIS

The topographical organization of differentiated cells into specialized tissues that is characteristic of early embryogenesis presents many examples of extensive cell migration and tissue remodeling. There are thus countless examples in which plasmino-gen activator may be involved.

In the analyses of early mouse embryos described above, it was apparent that more than one cell type was synthesizing plasmi-nogen activator. We have identified one of the cell types as parietal (distal) endoderm (Strickland *et al*, 1976), a type of extraembryonic endoderm that is responsible for the synthesis of the basement membrane known as Reichert's membrane. A variety of roles for plasminogen activator in parietal endoderm can be envisaged, e.g. it could be involved in the metabolism of Reichert's membrane, but no good evidence for a particular function has been reported.

Several previous publications suggested that in contrast to parietal endoderm, the other type of extraembryonic endoderm, visceral endoderm, does not synthesize plasminogen activator (Strickland *et al*, 1976; Jetten *et al*, 1979; Strickland & Sawey, 1980). One of these studies took special care to culture the visceral endoderm cells under optimum conditions (Jetten *et al*, 1979); another showed that although organ cultures of visceral yolk sacs produce copious amounts of alpha-fetoprotein, a visceral endoderm product, they synthesize no demonstrable plasminogen activator (Strickland & Sawey, 1980). However, Bode and Dziadek (1979) have reported that this protease is secreted by visceral endoderm, and also by mesoderm, and amnion. While the exact reasons for these discrepant observations are not yet known, several points should be noted in regard to the data of Bode and Dziadek. First, their experimental design does not allow the conclusion that the enzyme is actually being synthesized. Their experiments rely exclusively on the production of lytic zones in fibrin overlays by tissue fragments. Such experiments cannot distinguish between synthesis by the cells and simple adherent enzyme on the cells or extracellular matrix. Second, regardless of whether visceral endoderm can produce plasminogen activator under some conditions, their conclusion that this enzyme is not a useful marker to identify parietal endoderm is unwarranted. Under identical assay conditions in our hands and others, these two cell types differ profoundly in the amount of enzyme secreted.

The availability of such a marker for an early differentiated cell has made possible an analysis of the endocrine control of early development. Using a teratocarcinoma cell culture system as a model, it has been shown that retinoic acid at physiological concentrations can induce the differentiation of F9 embryonal carcinoma cells to endoderm (Strickland & Mahdavi, 1978). With the appropriate treatment (S. Strickland, K.K. Smith & K.R. Marotti, manuscript in preparation), the phenotype of the final cell leaves no doubt that it is definitive parietal endoderm.

6. CONCLUSIONS

The data summarized in this review are consistent with the view that plasminogen activator plays a general role in tissue remodeling. Such processes abound in morphogenesis, and the analysis of other specific examples should clarify how widespread this mechanism might be. This information may also have relevance to other situations, e.g. neoplasia, inflammation, in which tissue destruction occurs.

The versatility of the assays designed using enzyme production may prove eventually to be one of the most significant contributions from this line of work. The design of *in vitro* model systems to be used in the study of the hormonal control of cellular function requires a quantifiable property that can be correlated with the function. Plasminogen activator production provides one such property in some cases. This approach should help define the details of molecular endocrinology, and thereby facilitate rational intervention for the purposes of therapy or further investigations.

ACKNOWLEDGMENTS

The work from our laboratory described in this review was supported by grants from the American Cancer Society (CD-48), the National Institute of Child Health and Human Development (HD 11608), and The Rockefeller Foundation.

REFERENCES

AMOROSO, E.C. (1952) Placentation. In: A.S. Parkes (Ed.), Marshall's Physiology of Reproduction, Vol.II, Longmans, Green & Co., London-New York-Toronto, pp. 127-311.

ATWOOD, H.D. & PARK, W.W. (1961) J. Obstet. Gynaec. Brit. Comm. 68, 611.

BEERS, W.H. (1975) Follicular plasminogen and plasminogen activator and the effect of plasmin on ovarian follicle wall. Cell 6, 379-386.

BEERS, W.H., STRICKLAND, S. & REICH, E. (1975) Ovarian plasminogen activator: relationship to ovulation and hormonal regulation. Cell 6, 387-394.

BEERS, W.H. & STRICKLAND, S. (1978a) Involvement of plasminogen activator on ovulation. In: C.H. Spilman and J.W. Wilks (Eds.), Novel Aspects of Reproductive Physiology, Spectrum Publications, New York-London, pp. 13-26.

BEERS, W.H. & STRICKLAND, S. (1978b) A cell culture assay for follicle-stimulating hormone. J. Biol. Chem. 253, 3877-3881.

BODE, V.C. & DZIADEK, M.A. (1979) Plasminogen activator secretion during mouse embryogenesis. Develop. Biol. 73, 272-289.

BRAMBELL, F.W.R. (1956) Ovarian changes. In: A.S. Parkes (Ed.), Marshall's Physiology of Reproduction, Vol.I, Part 1, Longmans, Green & Co., London-New York-Toronto, pp. 397-542.

CARAVAGLIOS, R. & CILOTTI, R. (1957) A study of the proteins in the follicular fluid of the cow. J. Endocrinol. 15, 273-278.

CHRISTMAN, J.K., SILVERSTEIN, S.C. & ACS, G. (1977) Plasminogen activators. In: A.J. Barrett (Ed.), Proteinases in Mammalian Cells and Tissues, Elsevier/North Holland, New York, pp. 91-149.

DENKER, H.W. (1972) Blastocyst protease and implantation: effect of ovariectomy and progesterone substitution in the rabbit. Acta Endocrinol. 70, 591-602.

ESPEY, L.L. (1970) Effect of various substances on tensile strength of sow ovarian follicles. Amer. J. Physiol. 219, 230-233.

ESPEY, L.L. (1974) Ovarian proteolytic enzymes and ovulation. Biol. Reprod. 10, 216-235.

ESPEY, L.L. & LIPNER, H. (1965) Enzyme-induced rupture of rabbit Graafian follicle. Amer. J. Physiol. 208, 208-213.

DE GRAAF, R. (1672) De mulierum organis generationi inservientibus tractatus novus (English translation, J. Reprod. Fert., Suppl. 17, 77-107).

GROSS, J. (1974) Collagen biology: structure, degradation, and disease. The Harvey Lectures, Series 68, Academic Press, New York, pp. 351-432.

GROSS, J. & LAPIERE, C.M. (1962) Collagenolytic activity in amphibian tissues: a tissue culture assay. Proc. Natl. Acad. Sci. USA 48, 1014-1022.

GROSS, J. & NAGAI, Y. (1965) Specific degradation of the collagen molecule by tadpole collagenolytic enzyme. Proc. Natl. Acad. Sci. USA 54, 1197-1204.

HARRIS, E.D. & KRANE, S.M. (1974) Collagenases. N. Engl. J. Med. 291, 557-563.

JETTEN, A.M., JETTEN, M.E.R. & SHERMAN, M.I. (1979) Analyses of cell surface and secreted proteins of primary cultures of mouse extraembryonic membranes. Develop. Biol. 70, 89-104.

KAFATOS, F.C. (1972) The cocoonase zymogen cells of silk moths: a model of terminal cell differentiation for specific protein synthesis. In: A.A. Moscona and A. Monroy (Eds.), Current Topics in Developmental Biology, Academic Press, New York-London, pp. 125-191.

KIRBY, D.R.S. (1965) The "invasiveness" of the trophoblast. In: W.W. Park (Ed.), The Early Conceptus, Normal and Abnormal, University of St. Andrews Press, Edinburgh, pp. 68-73.

KIRCHNER, C. (1972) Uterine protease activities and lysis of the blastocyst coverings of the rabbit. J. Embryol. Exp. Morph. 28, 177-183.

LACROIX, M., SMITH, F.E. & FRITZ, I.B. (1977) Secretion of plasminogen activator by Sertoli cell enriched cultures. Mol. Cell. Endocrinol. 9, 227-236.

NOVAK, E.R. & WOODRUFF, J.D. (1974) Novak's Gynecologic and Obstetric Pathology, W.B. Saunders, Philadelphia, pp. 600.

MULLINS, D.E., BAZER, F.W. & ROBERTS, R.M. (1980) Secretion of a progesterone-induced inhibitor of plasminogen activator by the porcine uterus. Cell, in press.

OWERS, N.O. & BLANDAU, R.J. (1971) Proteolytic activity of the rat and guinea pig blastocyst in vitro. In: R.J. Blandau (Ed.), The Biology of the Blastocyst, The University of Chicago Press, Chicago, pp. 207-223.

PARR, E.L. (1974) Histological examination of the rat ovarian follicle wall prior to ovulation. Biol. Reprod. 11, 483-503.

PERLOFF, W.H., SCHULTZ, J., FARRIS, E.J. & BALIN, H. (1955) Some aspects of the chemical nature of human ovarian follicular fluid. Fertil. Steril. 6, 11-17.

PESEK, J. & BEERS, W.H. (1980) Plasminogen activator production by ovarian granulosa cells: induction in organ culture by luteinizing hormone. Biol. Reprod. in press.

PINSKER, M.C., SACCO, A.G. & MINTZ, B. (1974) Implantation associated proteinase in mouse uterine fluid. Develop. Biol. 38, 289-290.

ROMNEY, S.L., WHARTON, J.T. & FLETCHER, G. (1975) Choriocarcinoma. In: S.L. Romney, M.J. Gray, A.B. Little, J.A. Merrill, E.J. Quilligan, and R. Stander (Eds.), Gynecology and Obstetrics, McGraw-Hill, New York, pp.490.

SAKAI, T. & GROSS, J. (1967) Some properties of the products of the reaction of tadpole collagenase with collagen. Biochemistry 6, 518-528.

SAMUEL, C.A. (1971) The development of pig trophoblast in ectopic sites. J. Reprod. Fertil. 27, 494-495.

SAMUEL, C.A. & PERRY, J.S. (1972) The ultrastructure of the pig trophoblast transplanted to an ectopic site in the uterine wall. J. Anat. 113, 139-149.

SCHOCHET, S.S. (1916) A suggestion as to the process of ovulation and ovarian cyst formation. Anat. Rec. 10, 447-457.

SCHULTZE, H.E. & HEREMANS, J.F. (1966) Molecular Biology of Human Proteins. Elsevier, New York, pp. 219.

SHALGI, R., KRAICER, P., RIMON, A., PINTO, M. & SOFERMAN, N. (1973) Proteins of human follicular fluid: the blood-follicle barrier. Fertil. Steril. 24, 429-434.

SOLOMON, J.B. (1966) Relative growth of trophoblast and tumour cells co-implanted into isogenic mouse testis and the inhibitory action of "methotrexate". Nature 210, 716-718.

STRICKLAND, S. & BEERS, W.H. (1976) Studies on the role of plasminogen activator in ovulation: in vitro response of granulosa cells to gonado-trophins, cyclic nucleotides, and prostaglandins. J. Biol. Chem. 251, 5694-5702.

STRICKLAND, S., REICH, E. & SHERMAN, M.I. (1976) Plasminogen activator in early embryogenesis: enzyme production by trophoblast and parietal endoderm. Cell 9, 231-240.

STRICKLAND, S. & MAHDAVI, V. (1978) The induction of differentiation in teratocarcinoma stem cells by retinoic acid. Cell 15, 393-403.

STRICKLAND, S. & BEERS, W.H. (1979) Studies on the enzymatic basis and hormonal control of ovulation. In: A.R. Midgley and W.A. Sadler (Eds.), Ovarian Follicular Development and Function, Raven Press, New York, 143-153.

STRICKLAND, S. & SAWEY, M.J. (1980) Studies on the effect of retinoids on the differentiation of teratocarcinoma stem cells in vitro and in vivo. Develop. Biol., in press.

USUKU, G. & GROSS, J. (1965) Morphologic studies of connective tissue resorption in the tail fin of metamorphosing bullfrog tadpole. Develop. Biol. 11, 352-370.

WERB, Z., MAINARDI, C.L., VATER, C.A. & HARRIS, E.D. (1977) Endogenous activation of latent collagenase by rheumatoid synovial cells. N. Engl. J. Med. 296, 1017-1023.

© 1980 Elsevier/North-Holland Biomedical Press
Development in Mammals, Vol. 4, M.H. Johnson, editor

THE POLARIZING REGION AND LIMB DEVELOPMENT

C. Tickle

Department of Biology as Applied to Medicine,

The Middlesex Hospital Medical School,

London,

U.K.

1.   A SHORT OUTLINE OF LIMB DEVELOPMENT TO
INTRODUCE LIMB ANATOMY AND AXES

   Limbs develop in the flank of the embryo as a slight swelling,
which soon becomes more pronounced to form a definite bud. The
bud elongates to form a paddle-like structure. As the limb bud
continues to elongate, a kink develops which will become the
elbow or knee of the limb. Later a broad hand or foot plate can
be distinguished from which the digits are then carved. This
description of how the shape of the limb emerges could well apply
to both mammalian and bird limbs (Figure 1). Not only do the
limbs go through a similar series of outward morphological changes,
but it also turns out that the same basic principles underlie the
orderly laying down of the parts. The development of a 'paw' is
very like the development of a wing.
   A limb is a 3-dimensional structure, with several different
tissues, cartilage, bone, tendons, muscles and other soft tissues
arranged in a precise pattern. The pattern of tissues arises
from the pattern of cytodifferentiation of the mesenchyme cells
which make up the early bud. The problem then concerns the
spatial organization of cell differentiation and can be viewed
in terms of positional information (Wolpert, 1971). On this view,
pattern formation in the limb involves firstly assignment of
positional values to the mesenchyme cells, followed by interpretation leading to appropriate differentiation. One way in
which the position of a cell within the mesenchyme of the early
bud could be specified is in a 3-dimensional co-ordinate system
(Wolpert *et al*, 1975). The simplest co-ordinate system would
correspond to the three main axes of the limb: the proximo-
distal axis, which runs from the attachment of the limb at the
body to the tips of the digits, the antero-posterior axis, which
runs at right angles to the proximo-distal axis across the limb,
for example, in a hand from the thumb to the little finger, and
the remaining dorso-ventral axis, which runs, for example, from
the back of the hand to the palm. The pattern of skeletal
elements of a mouse fore-limb and a chick wing and the limb axes
are shown in Figure 2. It is with the development of the sequence
of structures across the antero-posterior axis that the polarizing
region is involved.

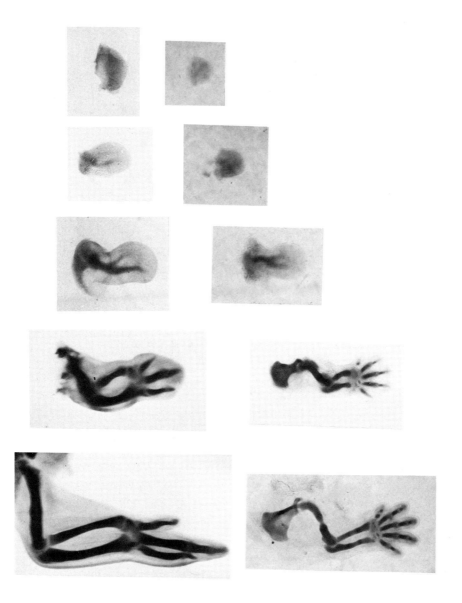

Figure 1: Stages in the development of limbs: whole mounts of chick wings (left) and mouse forelimbs (right) stained with alcian green to show cartilage. Chick limbs are staged according to Hamilton & Hamburger (1951) from top to bottom left: stages 20, 23, 26, 29, 33. Mouse limbs are staged according to day of gestation (day 0 = day of plug) from top to bottom right: stages 10, 10½, 12, 13, 14. Scale 1:8.

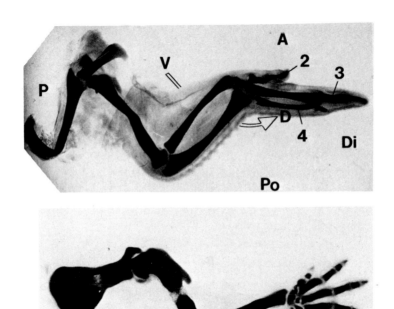

Figure 2: Dorsal view of whole mounts of right limbs stained with alcian green to show cartilage elements: chick wing (upper), mouse forelimb (lower). P = proximal, V = ventral, D = dorsal, Di = Distal, A = anterior, Po = Posterior; 2, 3 and 4 are digit numbers.

2.   HOW THE POLARIZING REGION OF THE CHICK WING WAS DISCOVERED

A major discovery about the control of pattern across the antero-posterior axis was made in the developing chick wing by Saunders and Gasseling (1968). A region at the posterior margin of the young bud was found to have a dramatic effect on the pattern across the antero-posterior axis of the wing if placed in an anterior position. The standard grafting procedure involved cutting a block of tissue out of the rim of the host bud and replacing it with the block of tissue from the posterior margin of a donor bud (Figure 3). The wing that resulted six days later had duplicated structures which were symmetrical about the mid-line (Figure 4). The transplanted region is called the polarizing region or zone of polarizing activity (ZPA). The discovery of the

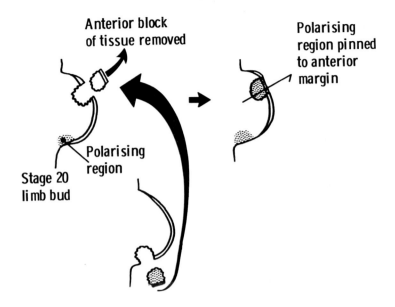

Figure 3: Diagram to show how standard graft to demonstrate polarizing activity is performed.

polarizing activity of the tissue at the posterior margin of the developing wing also accounts for the duplication of digits which results from grafting a wing tip rotated through 180° to its stump (Saunders & Gasseling, 1968). This operation brings anterior mesenchyme adjacent to the polarizing region.

The polarizing region, itself, does not give rise to the extra structures. These form from the adjacent mesenchyme under the influence of the grafted polarizing region. A particularly clear demonstration that the polarizing region is indeed a signalling region is observed in the result of grafting a polarizing region from a chick *leg* bud to a *wing* bud: extra *wing* digits are specified as with grafts of wing polarizing regions (Saunders & Gasseling, 1968; Summerbell & Tickle, 1977). Sometimes, however, the digit which forms next to the graft is a toe (Summerbell & Tickle, 1977) and is composed of both graft and host tissue.

3. THE LIMBS OF MICE AND OTHER MAMMALS HAVE A POLARIZING REGION TOO

Since similar grafting experiments on mammalian limbs have not so far been possible (see Section 10), the way in which a mammalian polarizing region was demonstrated involved grafting mammalian limb tissue to a developing chick wing bud. A region from the posterior margin of, for example, a hamster (MacCabe & Parker, 1976) or a mouse (Tickle et al, 1976) limb bud was found to effect a mirror-image duplication of a chick wing (Figure 5). The mouse or hamster polarizing region produces extra *wing* structures since the responding tissue is chick *wing* mesenchyme. These experiments not only showed that mammalian limbs have a polarizing region but also that the basis of the activity of the region is the same in both mammals and birds. Indeed, the limb buds of many other vertebrates, from snapping turtles to humans, have since been shown to have a polarizing region which acts in the chick wing (Fallon & Crosby, 1977). So, although all the experiments on how the polarizing region acts have been carried out with chick limbs, the information they provide is relevant to polarizing activity in the limbs of all vertebrates including mammals.

4. EFFECTS OF THE POLARIZING REGION

The effect of the polarizing region is most clearly seen in the duplicated pattern of wing digits. But the polarizing region affects the pattern of all wing elements which have not been laid down by the time the graft was made (Summerbell, 1974). Since the parts along the proximo-distal axis are laid down in sequence, proximal structures first, followed by progressively more distal ones, grafts to young buds affect more of the limb pattern than grafts to later buds. For example, the pattern of the fore-arm cartilage elements as well as that of the digits can be affected if the graft is made to a young bud in which only the upper arm has been laid down.

The pattern of the cartilage elements can be readily observed in whole mounted limbs as for example in Figures 2 and 4, and this provides a rapid way of analysing pattern. But the limb also contains other tissues arranged precisely. The patterns of muscles,

Figure 4: Dorsal view of chick wing that developed following a polarizing region graft to the anterior margin of the bud at stage 20. Digit pattern is 4 3 2 2 3 4 (compare with Figure 2 upper panel).

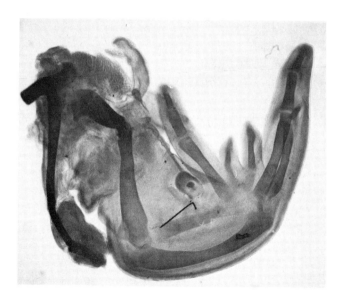

Figure 5: Dorsal view of chick wing that developed from a graft of tissue from the posterior margin of a mouse limb bud. Digit pattern is 3 2 2 3 4 (compare Figure 4).

tendons, blood vessels and nerves in limbs with a duplicated
pattern of cartilage elements are also duplicated (Shellswell &
Wolpert, 1977). One possibility is that the pattern of one tissue
determines the pattern of other tissues in the limb. This seems
to be the case for the nerves, which invade the limb well after
pattern formation is under way. In contrast, the patterns of
muscles, tendons and cartilage elements are more probably specified
independently, since muscles can be present in the absence of the
cartilage elements with which they are usually associated, and
tendons develop in the absence of the muscles to which they
normally attach (Shellswell & Wolpert, 1977). Thus, it appears
that the development of the pattern of muscles, tendons and
cartilage elements is independently controlled by the same system
of spatial cues from the polarizing region.

5. STRUCTURES ACROSS ANTERO-POSTERIOR AXIS OF THE LIMB DEVELOP ACCORDING TO THE DISTANCE FROM POLARIZING REGION

The relationship between the polarizing region and the pattern
of limb structures across the antero-posterior axis of the chick
wing has been analyzed by grafting a polarizing region to different positions around the margin of young buds (Tickle *et al*,
1975). It is simplest to consider the effects on the pattern of
the skeletal elements of the digits since each digit can be recognised; digit 4 being the most posterior digit, digit 2 the most
anterior, and digit 3 lying between digits 4 and 2 (Figure 2).
In the normal wing, the digits are formed from the posterior half
of the young bud and thus digit 2 develops about 0.5 mm from the
polarizing region. If an additional polarizing region is grafted
about 1 mm away from the host polarizing region, at the anterior
margin of the bud, the wing that develops possesses six digits in
mirror image symmetry, 4 3 2 2 3 4 (Figure 4). If the graft is
made slightly nearer the host polarizing region, the pattern of
digits that now results is 4 3 2 3 4 or even 4 3 3 4. If the
grafted polarizing region is placed more or less in the centre of
the limb bud rim, the wing that develops has a digit pattern of
2 3 4, 4 3 3 4; digits 2 3 and 4 arising from the tip of the wing
anterior to the polarizing region and digits 4 3 3 4 between the
graft and host polarizing regions (Figure 6). Thus, there is a

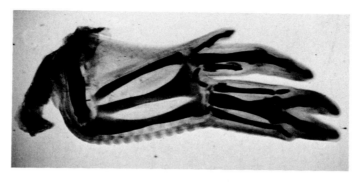

Figure 6: Dorsal view of chick wing that developed following a polarizing region graft to the centre of the bud margin at stage 20. Digit pattern is 2 3 (4), 4 3 3 4. Note that fore-arm contains three cartilage elements.

relationship between the distance of the tissue away from the polarizing region and which digit forms: digit 4 forming next to the polarizing region, digit 3 a bit farther away and digit 2 farther away still.

One idea of how distance from the polarizing region could be specified is that the polarizing region acts as a long range signaller, producing a diffusible morphogen which sets up a concentration gradient across the limb (Tickle *et al*, 1975). The concentration of morphogen at any point across the limb is a measure of the distance from the polarizing region and specifies which structures will form. Taking the digits as examples, digit 4, nearest the polarizing region, would be specified by a high concentration of morphogen, digit 2, farthest away, by a low concentration and digit 3, between digits 4 and 2, by an intermediate concentration. As the grafted polarizing region is placed more posteriorly, no digits 2 develop between graft and host polarizing region because the concentration of morphogen is too high. Indeed, if two additional polarizing regions are appropriately spaced around the rim of the bud, no digits 2 develop at all since no cells are far enough away from polarizing tissue (Figure 7).

The idea that the polarizing region signals by producing a diffusible morphogen has been discussed in some detail (Tickle *et al*, 1975). Both the time taken for the polarizing region to act and the distance over which the effect is exerted are compatible with diffusion, but as yet, a morphogen has not been discovered. Preliminary experiments which might pave the way to the isolation and

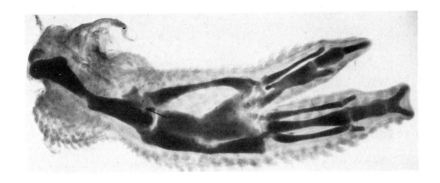

Figure 7: Dorsal view of chick wing that developed following two polarizing region grafts. One graft was to the anterior margin and the other was to the centre of the limb bud rim. Digit pattern is 4 3 4, 4 3 3 4. Note that 4 3 4 formed between the two grafted polarizing regions, and 4 3 3 4 between the more posterior graft and host polarizing region.

biochemical characterization of such a morphogen have, so far, been unsuccessful. For example, several kinds of extracts of polarizing region cells applied to limb buds on a range of inert carriers such as glass particles or ion exchange resins have no effect on the pattern across the antero-posterior axis (Section 6). One of the main pieces of evidence that the polarizing region acts at a distance, is that when the polarizing region is placed far anterior an additional digit 2, only, is produced. This result would be interpreted as being due to only the low part of the gradient lying within competent limb tissue.

Several other experimental results are consistent with this idea that the polarizing region acts by production of a diffusible morphogen. If the polarizing activity is weakened by, for example, very high doses of γ-radiation (Smith et al, 1978), and grafted to the anterior margin, digit 4 does not form next to the graft as usual. Instead, digit 3 is the digit specified nearest to the graft. With still more severe treatment an additional digit 2 only may be specified. With a gradient mechanism, it will always be most difficult to obtain the digit at the high point of the gradient (Summerbell & Tickle, 1977).

One of the first effects of a polarizing graft (at about 15h after grafting) is a widening of the limb. This increase in

width appears to be a primary effect of the graft and not,
trivially, due to the extra tissue of the additional digits, since
the widening occurs at the same time interval after grafting
irrespective of the stage of the development of the limb (Smith,
unpublished). It is not clear whether the increase in width is
due to a local proliferation of cells next to the graft or whether
a more extensive region of the wing is involved. In any case, an
accompaniment to the widening is the local persistence of the
specialised ectoderm overlying the wing mesenchyme, the so-called
apical ectodermal ridge. The persistence of the ridge permits the
outgrowth and laying down of distal structures from the anterior
part of the wing.

The widening of the limb is significant when plotting the concentration of a diffusible morphogen across the limb. If the
increase in width is taken into account, the pattern of digits
predicted from the concentration profiles fits the data well
(Figure 8). For example, if a polarizing region graft is placed
1 mm from the host polarizing region the concentration of
morphogen in the centre of the limb, assuming interaction between
the graft and host polarizing region, is too high to specify
digit 2. However, at the time the digits are being laid down,
the limb has widened by 50% and now the concentration in the
centre has fallen to specify digit 2. This result fits the experimental data well: when a polarizing region is placed 1 mm from the
host region, the pattern of digits is 4 3 2 2 3 4 or 4 3 2 3 4.
It is, therefore, of interest to examine the pattern of digits
obtained following grafts of polarizing region to buds which are
then prevented from widening. If distance is measured by the concentration of a diffusible morphogen, digit 2, farthest from the
polarizing region, should be lost since the concentration of
morphogen would remain relatively high throughout the narrow limb.
The results obtained fit this prediction. When widening of the
limb is inhibited either by reducing the number of cells in the
limb bud with X-irradiation (Smith, Hornbruch & Wolpert, unpublished) or by constraining the limb bud in a narrow tube
(Thorogood & Tickle, unpublished) digit(s) 2 is not specified.
The pattern of digits may even be just 4 3 4.

There is an alternative hypothesis to that of the diffusible
morphogen. It is inspired by the observed widening of the limb

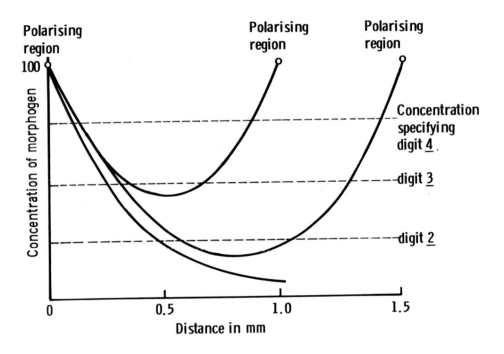

Figure 8: (above) Diagram to show how the zone of polarizing activity could specify the pattern of digits by production of a diffusible morphogen. In the normal limb (lower curve), it is assumed that the polarizing region is a source of morphogen which is kept at a concentration of 100. Since the morphogen is broken down as it diffuses across the limb, an exponential gradient is set up. The thresholds for specification of digits are shown. The posterior margin of the limb is at the left hand side of the diagram. If an additional polarizing region is grafted to the anterior margin, 1 mm away from the host polarizing region (upper curve), a U-shaped gradient results which would specify a digit pattern of 4 3 3 4. However, as the limb widens by 50% (middle curve), the concentration in the centre of the limb is now low enough to specify digit 2 and the pattern of digits is 4 3 2 2 3 4, a pattern which fits the experimental data well.

Figure 9: (opposite) Diagram to show how intercalation could account (b & c) for the pattern of digits obtained when a polarizing region is grafted to different positions along the antero-posterior axis and also (d) the pattern of digits predicted when two polarizing regions are grafted to the limb bud.

113

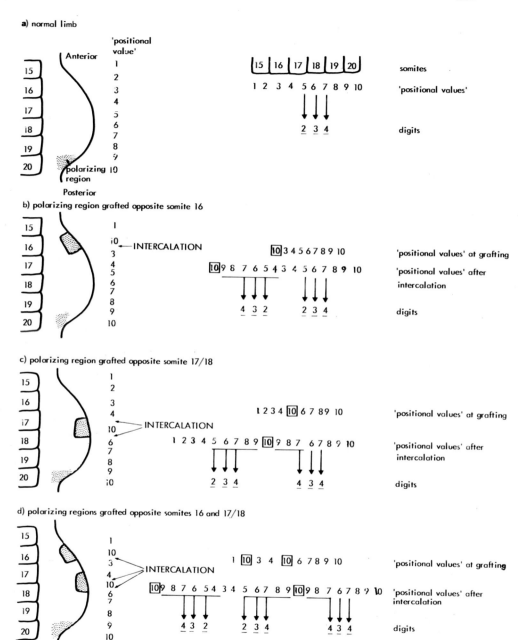

following a polarizing region graft, and proposes that the specification of additional wing structures (Iten & Murphy, 1980) is
analogous to the intercalary regeneration of cockroach and
amphibian limbs and insect imaginal discs (French *et al*, 1976).
Thus, the effects of a polarizing graft would involve local interaction between the polarizing region and adjacent tissue rather
than long range signalling. If developing chick limbs were
analogous to cockroach limbs, for example, then extra digits
should result from transplantation of any block of tissue from one
position along the antero-posterior axis to a different position
along this axis of the limb bud. The results so far (Iten &
Murphy, 1980) are difficult to interpret since it is not clear
whether the grafts, themselves, give rise to any structures.

Nevertheless, it is clear that intercalation could account for
the pattern of digits which follows grafts of the polarizing
region to different positions along the antero-posterior axis of
the limb bud. Thus, grafts made to the anterior margin would
result in filling a complete set of additional digits, while
grafts placed more posteriorly would fill in, between grafted and
host polarizing regions, a partial set of digits, a mirror image
of the structures which normally would develop posterior to the
graft, as well as producing a sequence of digits anterior to the
graft (Figure 9). A specific test of the intercalation idea is to
graft two polarizing regions to the bud, one at the anterior
margin and one at the centre of the limb bud rim. If all the
missing digits were to be filled in, it would be expected that a
complete duplicate set of digits would form between the two
grafted polarizing regions, since, normally, the anterior part of
the bud forms no digits at all. The results, however, show that
this is not the case: only a partial set of duplicated elements
usually arise, for example, 4 3 3 4 (Wolpert & Hornbruch,
unpublished) or 4 3 4 (Figure 7). This result instead shows the
relationship between the distance from the polarizing region and
which digits form. In addition, it is difficult to explain with
simple intercalation why, on the one hand, digit 2 is lost when
polarizing regions are placed close together (see also results
when widening is inhibited), while, on the other hand, digit 4 is
lost first when the polarizing region is weakened (Smith *et al*,

1978). These results also seem to rule out a sequential specification of digits since, again, there would be a conflict over
the order in which the digits are specified.

6. CELL BIOLOGY AND BIOCHEMISTRY OF THE POLARIZING REGION

Just how does the polarizing region exert its effect on pattern
formation? It is the mesenchyme cells and not the ectodermal
cells of the polarizing region that are involved in specifying
position across the antero-posterior axis (Saunders & Gasseling,
1968). These polarizing cells must be alive to have any effect.
Killing the polarizing region, by heating to 60°C for one hour or
by rapid freezing and thawing, destroys activity. However, controlled freezing of the polarizing region as a block of mesenchyme, which leaves some cells still viable, may preserve activity
(Tickle & Crawley, unpublished). This result is not surprising
since controlled freezing of suspensions of tissue culture cells
or chunks of transplantable tumours is routinely used to preserve
stocks of cells at particular points in their culture history
(Scherer & Hoogasian, 1954). Even very early mammalian embryos
can be preserved in this way (Whittingham et al, 1972). Thus, if
viable polarizing region cells are present, so is activity. This
technique may also provide a useful way of stockpiling polarizing
regions.

Polarizing regions treated with massive doses of γ-radiation
no longer signal, presumably because all the polarizing cells have
been killed. However, at progressively lower doses, polarizing
regions produce progressively more completely duplicated wing
patterns (Smith et al, 1978). A possible explanation is that more
polarizing cells remain viable as the dose of irradiation is
decreased.

A direct test of this idea has been made using cell reaggregates, in which the number of polarizing region cells has been
progressively diluted with non-signalling cells from the anterior
part of the wing (Tickle, Alberts & Goodman, unpublished). The
results show that there is indeed a quantitative relationship
between the number of polarizing region cells in the reaggregate
and the digit specified nearest the graft. Thus, with grafts of

reaggregated polarizing region cells, digit 4 is often specified
next to the graft, while with only 50% polarizing region cells
digit 4 is never formed. With still fewer polarizing region cells
in the graft, for example 20%, the most common result is that an
additional digit 2, only, is specified. Surprisingly, with as few
as 5% of the graft being from the polarizing region digit 2 is
still sometimes obtained. As in these experiments the graft is
the same size in each case, the number of polarizing cells required
to specify a particular digit next to the graft can be estimated
from the results. For digit 4, about 6,000 cells are required,
for digit 3, between 3,000 and 6,000 cells, and for digit 2, as
few as 300 cells may be sufficient. Thus, the number of grafted
polarizing region cells is related to which digit shall form next
to the graft. This conclusion is confirmed by the results of
grafts of mixtures of polarizing cells and hamster cells from the
cell line, BHK (Stoker & MacPherson, 1964). Again, as the number
of BHK cells used to make up the bulk of pellet is increased, the
digit formed next to graft becomes first digit 3 and eventually
digit 2 (Tickle & Goodman, unpublished).

A direct estimate of the number of cells required to specify
each digit can also be made by grafting small numbers of
polarizing region cells on inert carriers to the anterior margin
(Tickle & Alberts, unpublished). So far particles and small
pieces of melinex (200 μm square) have been used successfully.
The number of polarizing region cells which attach to a small
piece of melinex can be directly counted using a phase contrast
microscope. Estimation of the number of cells stuck to particles
can be made, in principle, using Hoescht dye to stain the nuclei
of the attached cells and counting them with a fluorescent microscope, although there are still some problems with this technique.
The present results show that as few as 100 cells grafted to a
wing on a piece of melinex may be sufficient to specify an
additional digit 4. The reasons for the difference in cell
numbers estimated in this way and those obtained from reaggregate
dilution experiments will be discussed later (Section 7).

Application of cells to wing buds on the inert carriers just
described, involved the preparation of cell suspensions from many
polarizing regions. A new technique, based on the use of filter

paper to make replicate cultures of cell clones enables small numbers of cells to be carried from the polarizing region of one wing to the anterior margin of another wing (Tickle & Alberts, unpublished). Slivers of paper, which have been inserted into slits in the polarizing region of one wing bud for a few hours can cause formation of an additional digit 2 when grafted to the anterior margin of another wing bud. It is indeed adherent polarizing region cells that are responsible for specification of the additional digit and not a 'morphogen' which has been sopped up by the paper, since rapid freezing and thawing of the paper prior to grafting destroys its polarizing activity.

Thus, several different lines of evidence show a quantitative relationship between the number of polarizing region cells and which digit forms next to the graft. This result is difficult to interpret with a simple intercalation model. In terms of the idea that the polarizing region produces a diffusible morphogen, it can be interpreted as fewer cells producing less morphogen. The attenuation observed suggests that there is no feed back to individual cells to produce more morphogen. In addition, there is no evidence for the neutral anterior cells being 'induced' to become polarizing region cells (see Section 8).

The attenuation of polarizing activity obtained when the polarizing region is treated with very high doses of γ-radiation (Smith *et al*, 1978) can be interpreted as being due to killing of polarizing cells. Grafting two such weakened polarizing regions together to the anterior margin of the wing has an additive effect (Smith, personal communication) and is consistent with this idea. The question is whether attenuation of signalling is always due to a reduced number of polarizing cells.

Attenuation of polarizing activity has also been found when polarizing regions have been subjected to an exhaustive range of chemical insults (Honing, Hornbruch, Smith & Wolpert, unpublished). The question is, therefore, whether this is due to non-specific cell damage or to a particular metabolic effect of the inhibitor. To answer this question, the biochemical activities of polarizing region cells were monitored at drug concentrations which both permit and prevent polarizing activity. From this series of experiments, several points emerge: drugs which inhibit oxidative phosphorylation have little effect on polarizing activity, whereas those which inhibit glycolysis abolish activity without affecting macromolecular synthesis. Inhibition of DNA synthesis has no effect on signalling. The effects of inhibition of protein and RNA synthesis are not so straightforward. While generally inhibition of protein and especially RNA synthesis reduces polarizing activity, nevertheless signalling can still occur with very low levels of protein and RNA synthesis.

As yet, therefore, there are few clues to the biochemical activities of the polarizing region cells necessary for signalling. There is even less insight into the nature or identity of any putative morphogen. Several hundred experiments have been carried out in which various extracts of polarizing regions have been applied to the anterior part of the wing on several types of inert carriers but with, as yet, no success (Tickle & Alberts, unpublished). A complementary approach of applying biologically active compounds, such as cyclic nucleotides, on a similar range of particles has also been unsuccessful.

Insulin was another of the compounds tested since this is the only substance, as far as we know, that has been found occasionally to lead to polydactyly when injected in ovo (Landauer, 1947). However, beads soaked in insulin, also had no effect on the pattern when grafted to developing chick wing buds. Nevertheless, this technology may ultimately prove to be a useful tool.

A positive aspect of these results is that additional digits do not form with just any non-specific insult. The polarizing region appears to be special. However, Saunders (1977) has recently found that tissues, other than the polarizing region, such as mesonephros and somites, can also cause additional digits to form if grafted to the anterior margin of a mesenchyme limb core, which is then re-combined with an ectodermal jacket. Our recent results confirm that mesonephros, for example, has low polarizing activity which can be detected if the graft is made to a slit under an intact apical ectodermal ridge (Tickle, Alberts & Goodman, unpublished). The basis for the low polarizing activity of these tissues remains obscure (see Section 7 and 8). The particular feature of the normal polarizing region appears to be the high activity of its constituent cells. It is surprising that as few

as 100 cells can specify digit 4. This small number of cells
suggests the possibility that the effect of the polarizing region
may be mediated by the cell surface.

7. INTERACTIONS BETWEEN THE POLARIZING REGION AND THE LIMB

The cells which can respond to the polarizing region are those
in the zone of undifferentiated mesenchyme at the tip of the
limb rimmed by the apical ectodermal ridge (Summerbell, 1974).
If a polarizing region is grafted to a more proximal position
along the margin of a later bud, the pattern of the resultant
wing is unaffected (Summerbell, 1974). To have any effect, the
polarizing region has to be placed in the region of the responding
cells.

The distribution of cells from a graft made to the anterior
margin of a young chick wing bud can be followed using quail
polarizing regions. Quail cells can be distinguished from chick
cells by a nucleolar marker, which can be easily recognized in
sections stained with the Feulgen technique (Le Douarin, 1973).
The quail cells taken from the polarizing region at the posterior
margin are found extending along the anterior margin and remain
near the tip of the limb during further out-growth (Summerbell &
Tickle, 1977). In contrast, quail cells from the anterior margin,
replacing a piece of chick tissue in the same position, do not
remain near the tip of the limb but are left in a proximal position
as the limb elongates. The different patterns of cell distribution
according to the origin of the graft may result from the different
response of the apical ectodermal ridge after each graft.

The propagation of grafted polarizing region cells in the zone
of undifferentiated mesenchyme throughout subsequent pattern formation suggests that polarizing cells may provide positional clues
as each 'segment' of the limb pattern is laid down in sequence.
Thus, the position of a cell across the antero-posterior axis of
the limb is specified by distance from the polarizing region,
while its position along the proximo-distal axis is specified by
the length of time spent in this zone of undifferentiated mesenchyme at the limb tip, the *progress zone* (Summerbell *et al*, 1973).
Cells leaving the progress zone would have their position along

the antero-posterior axis specified at the same time as their
position along the proximo-distal axis. Although this 'linked'
specification may occur during normal development, a polarizing
region graft does not have to remain in the progress zone to
affect the pattern of the most distal wing structures. A duplicate set of digits results following grafts of heavily irradiated
polarizing regions which are left behind as the limb grows out,
since polarizing cells so treated cannot divide (Smith, 1979).
Thus, cells in the progress zone remember exposure to the
polarizing region.

A grafted polarizing region not only has to be placed in the
progress zone to have any effect but also, more importantly, it
must be placed adjacent to the apical ectodermal ridge which rims
the tip of the limb. The relationship between the polarizing
region and the apical ridge is difficult to unravel, since
removal of part of the ridge prevents further limb outgrowth in
that part of the limb. As already mentioned (Section 5), an early
effect of a polarizing region graft to the anterior margin of a
bud is the widening of the bud accompanied by the maintenance and
thickening of the ridge over the anterior part of the bud.
Strangely, polarizing tissue itself causes immediately overlying
ridge to flatten (Saunders & Gasseling, 1968). The effect of the
polarizing region on apical ridge thickening therefore appears to
be at a distance. A crucial question is whether this is a primary
effect of the polarizing region or a secondary effect which results
from re-specification of the mesenchyme.

It seems likely that the effect on the ridge is a secondary one,
since the height of the ridge always conforms to the underlying
mesenchyme and is reversible. For example, if the ectodermal
covering of a normal limb bud is rotated 180° relative to the
mesenchyme core, the posterior thickening of the ridge now
diminishes when placed in contact with anterior mesenchyme and
the anterior part of the ridge, next to posterior mesenchyme
thickens (Zwilling, 1956). This conclusion is confirmed by
analysis of the ability of the ridge to 'remember' exposure to
the polarizing region which shows that the effect on the ridge is
reversible and conforms to the polarity of the mesenchyme core
with which it is combined (MacCabe & Parker, 1979). The effect of

polarizing tissue on the thickening of the apical ridge has provided the basis for a rapid *in vitro* assay for polarizing
activity (MacCabe & Parker, 1975). The assay depends on the
effects of culturing small portions of the anterior tip of a limb
bud with and without polarizing tissue for 48 hours. In the
absence of polarizing tissue, the apical ridge flattens and many
macrophages appear in the mesenchyme. Ridge flattening is
inhibited and macrophages do not appear when polarizing tissue is
added to the culture. This ability to maintain the ridge and
inhibit cell death is not confined just to the polarizing region,
but is exhibited in a graded form by tissue across the posterior
half of the limb bud. It seems likely that the graded property
of the mesenchyme reflects the effects of the polarizing region
rather than the concentration of a diffusible 'morphogen'.
Nevertheless, some progress has been made towards identifying the
factor involved and cell-free extracts have been prepared which
mimic the effect of polarizing tissue (Calandra & MacCable, 1978).

The spatial relationship between the ridge and the polarizing
region appears to be important for signalling. Tissues, which
previously have been considered to be inert, can be shown to
possess activity if grafted to a mesenchyme core, which is then
covered by an ectodermal jacket (Saunders, 1977). A possible
interpretation is that lower polarizing activities can be detected
if polarizing cells are placed in contact with an unbroken apical
ridge. The quantitative relationship between the number of
polarizing cells and the digit which forms adjacent to the graft
provides a measure for signalling efficiency and can be used to
test whether grafts made under an intact apical ridge are indeed
more effective than those made in the standard way which involves
removing part of the ridge (Figure 3). Using a simpler technique
than that of Saunders (1977) which still leaves the ridge unbroken
(Figure 10), it is clear that cells placed under an intact ridge
are more effective, in terms of the digit specified next to the
graft, than those made in the standard way (Tickle *et al*,
unpublished). If a pellet of polarizing region cells, containing
approximately 6,000 cells, is placed into a slit under the ridge
at the anterior edge of a bud, digit 4 is invariably formed next
to the graft. In the normal grafts, digit 4 forms in 40% of the

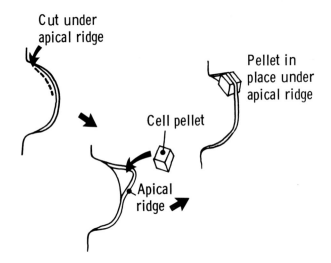

Figure 10: Diagram to show how polarizing activity is assayed by grafting tissue to a slit under an intact apical ridge. (Compare with Figure 3).

cases. Again, with 50% of polarizing cells in a pellet (3,000 cells) placed under an unbroken ridge,digit 4 still forms in some cases while such a result is never obtained with the standard graft. The signalling efficiency of grafts under the ridge is compared with that of standard grafts in Figure 11. In addition, with this grafting technique, weak polarizing activity can be detected in mesonephros, confirming Saunders result (1977).

A possible explanation for increased efficiency of signalling of grafts under an intact apical ridge is that only those polarizing cells which come to lie directly beneath the ridge are involved in specification of additional structures. This idea may explain the observation that reaggregates made from cells which have been freed from their neighbours, are more effective grafts than pieces of intact tissue (Saunders, 1977). Known numbers of polarizing cells can be placed directly beneath intact ridges on small pieces of plastic. As few as 100 polarizing cells can specify digit 4 next to the graft, and approximately 50 cells can specify digit 3 (Tickle et al, unpublished). These results suggest that very few polarizing cells are required to specify additional digits if presented as a cell monolayer under the ridge. Thus, in grafts which contain around 6,000 cells, it is possible that only the cell layer which underlies the ridge is important for specifying additional digits, the bulk of the graft containing non-signalling or ineffective cells. This explanation fits the data obtained so far remarkably well. Table 1 compares the number of cells on pieces of plastic which are required to specify successive digits with estimates of the number of polarizing cells on the face of a pellet which specifies the same digit. The data are consistent with the idea that a single layer of cells is required, the rest of the graft being superfluous.

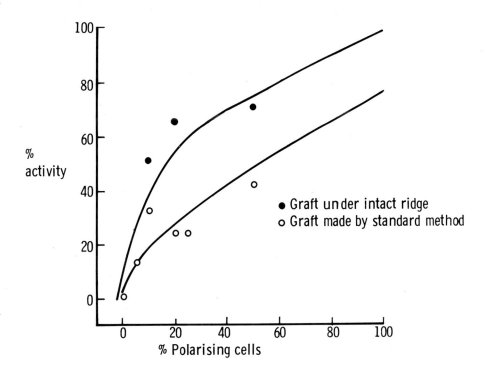

Figure 11: Diagram to show difference in effectiveness of grafts made by standard method and by insertion into a slit beneath an intact apical ridge. The effectiveness of the graft is represented by % polarizing activity: this is calculated by giving scores for the digit specified next to graft, digit 4 scores 3, digit 3 scores 2, digit 2 scores 1 and a knob scores ½, and expressing total score as a percentage of total possible score.

TABLE 1: COMPARISON OF INDEPENDENT ESTIMATES OF THE
NUMBER OF POLARIZING CELLS, PLACED UNDER AN
INTACT APICAL RIDGE, WHICH SPECIFY EACH
ADDITIONAL WING DIGIT

	Digit 4	Digit 3	Digit 2
a) Number of polarizing cells attached to small pieces of melinex*	100	40 50	30 80
b) Number of polarizing cells on a face of a piece of pellet of reaggregated polarizing and anterior mesenchyme cells†	150	30-60	30 or less

* Data refer to individual experiments

†Data calculated from average number of cells
on face of a pellet and % polarizing cells in
mixed pellet required to specify each digit.

Using monolayers of cells on inert carriers, such as small
pieces of plastic, enables the relationship between ridge and
polarizing region cells to be examined further. The surface
carrying the cells can be implanted either facing the ridge or
away from the ridge. Preliminary results are not open to clear
cut interpretation: additional digits are specified with the cells
facing either way. Possibly the cells migrate from the plastic
into position under the ridge and this will be examined using
quail cells.

An interesting possibility, arising from the enhanced effectiveness of polarizing region cells placed under the ridge, is that
signalling to the adjacent mesenchyme cells which form extra
digits takes place via the ridge. In principle, the effect of
the polarizing region could travel either through the mesenchyme
or through the apical ridge. There are problems associated with
both possible routes.

It is not clear whether cell to cell contact between polarizing cells and responding mesenchyme cells is required. Barrier experiments by Saunders and Gasseling (1963) appear to show that activity can be transmitted through millipore filters with very small pores, which would exclude cell to cell contact. However, nucleopore barriers, with even quite large holes inserted between a polarizing graft and responding mesenchyme prevent the formation of additional digits (Summerbell & Tickle, unpublished). It should be noted that in these experiments, the ridge as well as the mesenchyme was interrupted by the barrier. Nevertheless, it would seem that passage of the signal via the mesenchyme would need to be from cell to cell since extracellular diffusion would lead to dilution in the numerous spaces between cells and in blood vessels. Indeed, the mesenchyme cells are linked by gap junctions (Kelley & Fallon, 1978) and limb mesenchyme cells can metabolically communicate in culture with BHK cells (Pitts & Tickle, unpublished) and with each other (Morris, personal communication). In addition, mixing polarizing cells with non-communicating L cells (Gilula et al, 1972) reduces the signalling efficiency as assessed by which digit forms next to the graft (Tickle & Goodman, in preparation). However, the limb mesenchyme cells in vivo contact their neighbours over such a small proportion of their surface that diffusion may be limited and not sufficiently rapid to set up a concentration gradient in the time required. With diffusion through the ridge this would not be a problem since the ridge cells are linked by extensive gap junctions (Fallon & Kelley, 1977). With an ectodermal route the problem is how the signal passes from the polarizing region cells into the apical ridge since a basal lamina intervenes.

To what extent the basal lamina acts as a barrier is not known. Preliminary experiments show that the basal lamina is permeable to molecules such as horseradish peroxidase which, when applied externally, can seep through the ectoderm into the underlying mesenchyme. The ectoderm and ridge do not appear to be isolated from the mesenchyme by the basal lamina. Indeed, there are report that the basal lamina under the ridge is incomplete (reviewed by Ede, 1971), although no direct cell to cell contacts between mesenchyme and apical ridge cells have been observed. However, chick limb ectoderm cells and mesenchyme cells do not appear to be able to communicate in culture, as judged by the passage of small molecules (Pitts & Simms, 1977) even in the absence of a basal lamina (J. Morris, personal communication). Communication between mesenchyme and ridge cells is now being examined.

8. THE ROLE OF THE POLARIZING REGION IN NORMAL DEVELOPMENT INCLUDING ITS ORIGIN

In amphibian embryos, well before the limbs appear, narrow strips of flank tissue, just posterior to the level at which limbs will subsequently develop, will effect duplication of limb structures when placed into anterior positions and act as polarizing regions (Slack, 1976). An interesting speculation is that the polarizing region for both hind and fore limbs arises from small groups of mesenchyme cells that pass through the blastopore lip at specific times during gastrulation. If this idea was also applicable to the origin of the polarizing region of bird limbs, it might explain why other mesodermally derived tissues, such as somites and mesonephros, at the level of the limbs have also been found to have polarizing activity (Saunders, 1977). A similar series of grafting experiments, involving small strips of flank tissue does not appear to have been carried out in chick embryos. However, the antero-posterior axis of pre-sumptive limb-forming mesenchyme does appear to be fixed before the limbs appear in chick embryos. Grafts of such pieces of tissue to varying positions along the flank develop into limbs with axes corresponding with the origin of the tissue and not with the site of implantation (Saunders & Reuss, 1974).

Once limb buds are formed and during outgrowth from the body wall, polarizing activity can clearly be demonstrated in tissue along the posterior margin of the bud by grafting experiments. An exhaustive series of grafts have enabled maps of polarizing activity in chick limbs of different stages to be drawn (MacCabe et al, 1973). The polarizing region can thus be defined by its position in the developing limb. No morphological characteristics have been found which allow identification of polarizing cells.

A striking feature of maps of polarizing activity is that activity is located in progressively more distal tissue as the limb elongates, remaining at the edge of the progress zone. One possibility is that polarizing tissue induces adjacent distal tissue to become polarizing tissue (Saunders, 1972). This is, however, ruled out by the absence of the polarizing activity in tissue distal to grafts of heavily irradiated polarizing regions which do not divide and propagate along the anterior margin following grafting (Smith, 1979). This result suggests instead that the polarizing region is heritably propagated from a group of cells at the posterior tip of the developing limb (Summerbell & Tickle, 1977). From the maps of polarizing activity, it is apparent that polarizing tissue occupies the same region in the early bud as the posterior necrotic zone. However, the extensive cell death in this region is not the basis of polarizing activity (Saunders & Gasseling, 1968). A further interesting feature of maps of polarizing activity, in view of a monolayer of polarizing cells specifying additional digits, is that polarizing activity is not confined to a line of cells at the posterior margin - indeed, the small numbers of cells from polarizing regions which are seeded on to small sheets of plastic as grafts, make it extremely unlikely that only the single layer of cells at the posterior edge of the wing is active. Nevertheless, during normal development it is possible that only this layer of cells is important.

A question that is often asked about the polarizing region is what happens when it is cut out. A problem is that it is very difficult to be certain that all cells with polarizing activity have been removed. Removal of tissue in which polarizing activity has been detected results in more or less normal wings (MacCabe et al, 1973). However, other workers removing similar pieces of polarizing tissue have found that activity can be assayed from the posterior margin of the host limb after healing (Fallon & Crosby, 1975). Nevertheless, if sufficient of the posterior margin of a developing chick wing is cut away so that not only polarizing tissue but also tissue which would form digit 4 is removed, digits 3 and 2 develop normally (Tickle et al, 1975). If all the polarizing tissue has been removed, the subsequent development of digits 2 and 3 can be interpreted as showing that cells can remember exposure to the polarizing region: in terms of a diffusible morphogen, the cells differentiate according to the highest exposure ever experienced (Lewis *et al*, 1977; Meinhardt, 1978). An analogy would be with a ratchet, once the cells have clicked, they cannot reach a lower position. As already mentioned (Section 7) evidence for a positional memory following grafts of heavily irradiated polarizing region cells is that distal structures are still duplicated even though the graft does not remain in the progress zone (Smith, 1979). In addition, such heavily irradiated polarizing regions can be readily removed after 15 hours and the host limb still develops duplicated distal structures (Smith, unpublished). A difficulty with the idea that cells differentiate according to the highest concentration of morphogen ever experienced and have a positional memory is the apparent creation of more anterior structures distally in duplicated limbs. For example, digit 3 can branch into two digit 3s distally (Figure 6). This phenomenon is also clearly shown in the patterns of duplicated amphibian limbs, which result from grafts made at very early stages (Slack, 1977) and has been termed 'distal deepening'. For example, consistent rules emerge about the relationship of the number of fore-arm elements and the number of digits in duplicated limbs. A similar relationship has now been established for the patterns of elements in duplicated chick wings (Tickle *et al*, unpublished). Distal deepening and long-term positional memory are difficult to reconcile.

The results of recent experiments, in which impermeable barriers are placed across the bud so as to divide it into anterior and posterior parts suggests, however, that positional memory may be rather short-lived (Summerbell, 1979). The pattern of structures which form when the barrier is placed in successive positions along the antero-posterior axis of the bud cannot be fitted to a simple fate map. Tissue anterior to the barrier and isolated from the polarizing region forms fewer structures than would be expected. These results fit simulations of the gradients of diffusible morphogen that would result from insertion of impermeable barriers and can be interpreted as showing that the polarizing region is active during normal development.

9.   POLYDACTYLOUS LIMB MUTANTS OF CHICKS PROVIDE FEW
     INSIGHTS INTO THE ACTION OF THE POLARIZING REGION

There are some mutants of chicks in which the limbs possess
extra digits. In some mutants, for example, talpid$_2$ (Abbot et al,
1960) and talpid$_3$ (Hinchliffe & Ede, 1967), many repeated cartilage elements develop from a broad hand plate but cannot be identified as specific digits. In other mutants, however, such as duplicate polydactyly (Zwilling & Hansborough, 1956) and diplopodia$_4$
(Abbot & Kieny, 1961) about twice the normal number of digits
develop and most of these can be identified. In diplopodia$_4$, the
posterior digits correspond to those of a normal limb while those
anteriorly are often copies of the most anterior digit.

A common feature of all these polydactylous mutant limbs is
that the bud from early stages is wider along the antero-posterior
axis than normal limb buds. The production of additional digits
from the anterior part of the bud seems to be correlated with the
maintenance of the apical ridge over this part of the developing
limb. Whether it is the ectoderm or the mesenchyme of the limb
that is defective in these mutants can be tested by separating
the two components and combining with normal limb tissue. For
example, mutant ectodermal hulls can be recombined with normal
mesenchymal cores. These experiments have been performed for a
number of polydactylous mutants and in each case, it is the
mesenchyme that is defective, not the ectoderm (reviewed by
Wolpert, 1976).

The increased width of mutant limb buds is reminiscent of the
widening of a normal bud following a polarizing region graft to
the anterior margin, which also results in additional digits. The
activity of polarizing cells of diplopodia$_4$ limb buds has been
assayed by grafts to normal buds. The results show that the
mutant polarizing region is indistinguishable in its effect from
a normal polarizing region (MacCabe et al, 1975). This result
rules out the possibility that the extra digits form in the mutant
because of enhanced polarizing activity. A second possibility,
that has not been investigated, is that there is weak polarizing
activity at the anterior margin of the mutant bud. A few
polarizing cells would be sufficient to specify additional digit 2s.
Nevertheless, it seems probable that in polydactylous mutants it

is the response of the mesenchyme cells to the signal that is
changed and not the intensity or distribution of the signal
itself.

10. DEVELOPMENT OF STRUCTURES ACROSS THE ANTERO-POSTERIOR AXIS OF THE MOUSE LIMB

10.1 NEW TECHNIQUES FOR CULTURING MOUSE EMBRYOS MAY ALLOW GRAFTING EXPERIMENTS

The development of structures across the antero-posterior
axis of the mouse limb follows the same rules as outlined for the
chick limb. This can be inferred from the effect of tissue from
the posterior margin of a mouse limb bud: extra structures are
specified in the chick wing. The development of five digits of a
mouse paw as opposed to the development of three digits in a chick
wing results from the different interpretation of the signal from
the polarizing region by the mouse limb mesenchyme cells.

Grafting experiments on developing mammalian limbs may soon be
feasible. Rat embryos can be explanted at early limb bud stages
and grown in culture until the digits develop (Cockcroft, 1973;
also reviewed by New, 1978). Indeed, pioneering experiments, in
which simple operations were performed on developing rat limbs,
have been carried out successfully with embryos in culture
(Deuchar, 1976). At present, however, these techniques are in
their infancy.

Another possible approach, which would allow operations to be
performed on developing mammalian limbs, is to culture the limb
buds themselves rather than the whole embryo. Mammalian limb buds
develop reasonably well in explant culture (Lessmöllmann et al,
1975) although the skeleton is sometimes stunted and the joints
fused. It appears that no-one has tried grafting experiments with
buds in such organ cultures. Rather than explanting the bud in
organ culture, the possibilities of in vivo culture also exist,
on, for example, the chorio-allantoic membrane of the chick.
However, chick limb buds which have developed on the chorio-
allantoic membrane are, as are mouse limbs in organ culture,
rather stunted. Another possibility is to graft the mouse limb
bud in place of the chick wing bud in ovo, since a replacement
chick wing or even leg develops normally in this site. However,
it has not been possible to grow whole mouse limbs on chick wing

bud stumps for any length of time, even though small pieces of
mouse limb mesenchyme grow and differentiate when grafted to a
chick wing (Cairns, 1965).

10.2 THE DEVELOPING CHICK LIMB ALLOWS INVESTIGATION OF THE BASIS OF POLARIZING ACTIVITY OF MOUSE LIMBS TOO

Despite the difficulties of performing direct grafting experiments on developing mammalian limbs, the properties of the
mammalian polarizing region can be investigated in the chick wing.
It is possible, for example, to investigate the effects of various
treatments of the mammalian polarizing region by assaying its
activity in producing additional wing structures. In the same
way, estimates of the number of mammalian polarizing cells
required to specify particular wing digits can be made.

An intriguing feature of the duplicated wings which result
following grafts of mammalian polarizing regions is that digit 4
is rarely, if ever, specified next to the graft (Tickle *et al*,
1976; MacCabe & Parker, 1976; Fallon & Crosby, 1977, and Figure 5).
Whether this is due to death of some of the cells in the grafts
or reduced transmission of the signal is unknown.

Implantation to the developing chick wing bud thus provides a
way of investigating the activity of mammalian polarizing regions.
Mammalian cells may, themselves, provide additional scope for
investigating the basis of polarizing activity. For example, the
antigens of cells from different strains of mice are well
characterized and antibodies are available, whereas this is not
the case for chick cells. The effect of blocking specific cell
surface components on signalling of mouse polarizing cells could
be explored.

Another possible area in which mouse cells may offer advantages
over chick cells is that of cell culture since culture techniques
have been well worked out for mammalian cells and continuous cell
lines have been derived. It is doubtful whether any continuous
chick cell lines have ever been established. So far, a frustrating
property of chick polarizing cells is loss of activity on culturing, either as solid pieces of tissue (Saunders, 1972) or as
cell monolayers. The effect of culture on mouse polarizing cells
is worth investigating since, if large numbers of polarizing cells

could be obtained, biochemical investigations would be much more
straightforward.

There are several different mouse mutants in which polydactyly
has been described (see Green, 1966). In some cases, the hind
limb only is affected whereas in others, both fore and hind limbs
have extra digits. The basis for the production of the additional
digits in these mutants remains to be explored, but it seems probable, by analogy with the polydactylous chick mutants, that they
arise due to a changed interpretation of positional information
from the polarizing region.

A curious finding is that some cytotoxic drugs, such as
cytosine arabinoside, lead to the formation of additional digits
when applied to developing mammalian embryos (Scott et al, 1975,
1977). The most usual result is the development of a copy of the
most anterior digit. The development of the additional digit
appears to be correlated with the absence of the cell degeneration
which normally occurs at the anterior margin of the limb.

11. SUMMING UP

An apology for discussing so many experiments on the developing
chick wing in a series on mammalian development should be
unnecessary. It seems reasonable to conclude that the chick wing
and mammalian limb develop in much the same way. The position of
a cell across the antero-posterior axis of the limb appears to be
specified by distance from a polarizing region, which has been
identified in both bird and mammalian limbs. In addition,
position along the proximo-distal axis is probably specified in
the same way since the apical ridge of a mouse limb can permit the
laying down of the structures of a chick wing along this axis
(Jorquera & Pugin, 1971).

The problem of how the polarizing region acts to specify
position along the antero-posterior axis of a limb remains a
fascinating one. However, there is a growing body of information
about its action. It is encouraging that a start has been made in
understanding the cell biology of polarizing cells. This start
will lead to new avenues of research away from the traditional

cut-and-paste approach which has limitations. Indeed, the future offers exciting possibilities if advantage is taken of the gamut of innovative techniques in cell biology. For example, manipulations of both the cell surface and of the composition of cell contents, which have been worked out for tissue culture cells, can be exploited. It is here that mammalian polarizing regions may be important tools. Thus, the basis of polarizing activity may be eventually understood at the cellular level and ultimately at the biochemical level, too.

Finally, the study of the polarizing region, besides being intrinsically interesting, may, in addition, provide insights into other developmental signalling regions.

ACKNOWLEDGEMENTS

My unpublished work is supported by the M.R.C. I thank Professor L. Wolpert for his advice and encouragement and A. Crawley for preparing some of the figures.

REFERENCES

ABBOTT, U. & KIENY, M. (1961) Sur la croissance in vitro du tibiotarse et du perone de l'embryon de poulet "diplopode". Compt. Rend. Acad. Sci. Paris 252, 1863-1865.

ABBOTT, U.K., TAYLOR, L.W. & ABPLANALP, H. (1960) Studies with talpid$_2$, an embryonic lethal of the fowl. J. Heredity 51, 195-202.

CAIRNS, J.M. (1965) Development of grafts from mouse embryos to the wing bud of the chick embryo. Dev. Biol. 12, 36-52.

CALANDRA, A.J. & MacCABE, J.A. (1978) The in vitro maintenance of the limb-bud apical ridge by cell-free preparations. Dev. Biol. 62, 258-269.

COCKCROFT, D.L. (1973) Development in culture of rat foetuses explanted at 12.5 and 13.5 days of gestation. J. Embryol. exp. Morph. 29, 473-483.

DEUCHAR, E.M. (1976) Regeneration of amputated limb-buds in early rat embryos. J. Embryol. exp. Morph. 35, 345-354.

EDE, D.A. (1971) Control of form and pattern in the vertebrate limb. In: D.D. Davies and M. Balls (Eds.), Control Mechanisms of Growth and Differentiation, Cambridge University Press, Cambridge, pp. 235-254.

FALLON, J.F. & CROSBY, G.M. (1975) Normal development of the chick wing following removal of the polarizing zone. J. exp. Zool. 193, 449-455.

FALLON, J.F. & CROSBY, G.M. (1977) Polarizing zone activity in limb buds of amniotes. In: D.A. Ede, J.R. Hinchliffe and M. Balls (Eds.), Vertebrate Limb and Somite Morphogenesis, Cambridge University Press, Cambridge, pp.55-71.

FALLON, J.F. & KELLEY, R.O. (1977) Ultrastructural analysis of the apical ectodermal ridge during vertebrate limb morphogenesis. J. Embryol. exp. Morp. 41, 223-232.

FRENCH, V., BRYANT, P.J. & BRYANT, S.V. (1976) Pattern regulation in epimorphic fields. Science, N.Y. 193, 969-981.

GILULA, N.B., REEVES, O.R. & STEINBACH, A. (1972) Metabolic coupling, ionic coupling and cell contacts. Nature, Lond. 235, 262-265.

GREEN, M.C. (1966) Mutant genes and linkages. In: E.L. Green (Ed.) Biology of the Laboratory Mouse, 2nd Edn., McGraw-Hill, New York, pp. 87-151.

HAMBURGER, V. & HAMILTON, H.L. (1951) A series of normal stages in development of the chick embryo. J. Morph. 88, 49-92.

HINCHLIFFE, J.R. & EDE, D.A. (1967) Limb development in the polydactylous talpid$_3$ mutant of the fowl. J. Embryol. exp. Morph. 17, 385-404.

ITEN, L. & MURPHY, L. (1980) Dev. Biol. in press.

JORQUERA, B. & PUGIN, E. (1971) Sur le comportement du mesoderme et de l'ectoderme du bourgeon de membre dans les echanges entre le poulet et le rat. Compt. Rend. Acad. Sci. Paris 272, 1522-1525.

KELLEY, R.O. & FALLON, J.F. (1978) Identification and distribution of gap junctions in the mesoderm of the developing chick limb bud. J. Embryol. exp. Morph. 46, 99-110.

LANDAUER, W. (1947) Insulin-induced abnormalities of beak, extremities and eyes of chickens. J. exp. Zool. 105, 145-172.

LE DOUARIN, N. (1973) A biological cell labelling technique and its use in experimental embryology. Dev. Biol. 30, 217-222.

LESSMOLLMANN, U., NEUBERT, D. & MERKER, H-J (1975) Mammalian limb buds differentiating in vitro as a test system for the evaluation of embryotoxic effects. In: D. Neubert and H-J. Merker (Eds.) New Approaches to the Evaluation of Abnormal Embryonic Development, Georg Thieme: Stuttgart, pp. 99-113.

LEWIS, J.H., SLACK, J.M.W. & WOLPERT, L. (1977) Thresholds in development. J. theor. Biol. 65, 549-590.

MacCABE, A.B., GASSELING, M.T. & SAUNDERS, J.W. (1973) Spatiotemporal distribution of mechanisms that control outgrowth and anteroposterior polarization of the limb bud in the chick embryo. Mech. Ageing Devl. 2, 1-12.

MacCABE, J.A., MacCABE, A.B., ABBOTT, U.K. & McCARREY, J.R. (1975) Limb development in diplopodia$_4$: a polydactylous mutation in the chicken. J. exp. Zool. 191, 383-394.

MacCABE, J.A. & PARKER, B.W. (1975) The in vitro maintenance of the apical ectodermal ridge of the chick embryo wing bud: an assay for polarizing activity. Dev. Biol. 45, 349-357.

MacCABE, J.A. & PARKER, B.W. (1976) Polarizing activity in the developing limb of the syrian hamster. J. exp. Zool. 195, 311-317.

MacCABE, J.A. & PARKER, B.W. (1979) The target tissue of limb-bud polarizing activity in the induction of supernumerary structures. J. Embryol. exp. Morph. 53, 67-73.

MEINHARDT, H. (1978) Space-dependent cell determination under the control of a morphogen gradient. J. theor. Biol. 74, 307-321.

NEW, D.A.T. (1978) Whole-embryo culture and the study of mammalian embryos during organogenesis. Biol. Rev. 53, 81-122.

PITTS, J.D. & SIMMS, J.W. (1977) Permeability of junctions between animal cells. Intercellular transfer of nucleotides but not of macromolecules. Exp. Cell Res. 104, 153-163.

SAUNDERS, J.W. (1972) Developmental control of three-dimensional polarity in the avian limb. Ann. N.Y. Acad. Sci. 193, 29-42.

SAUNDERS, J.W. (1977) The experimental analysis of chick limb bud development. In: D.A. Ede, J.R. Hinchliffe and M. Balls (Eds.), Vertebrate Limb and Somite Morphogenesis, Cambridge University Press, Cambridge, pp. 1-24.

SAUNDERS, J.W. & GASSELING, M.T. (1963) Trans-filter propagation of apical ectoderm maintenance factor in the chick embryo wing bud. Dev. Biol. 7, 64-78.

SAUNDERS, J.W. & GASSELING, M.T. (1968) Ectodermal-mesodermal interactions in the origin of limb symmetry. In: R. Fleischmajer and R.E. Billingham (Eds.), Epithelial-mesenchymal interactions, Williams and Wilkins: Baltimore, pp. 78-97.

SAUNDERS, J.W. & REUSS, C. (1974) Inductive and axial properties of prospective wing bud mesoderm in the chick embryo. Dev. Biol. 38, 41-50.

SCHERER, W.F. & HOOGASIAN, A.C. (1954) Preservation at subzero temperatures of mouse fibroblasts (Strain L) and human epithelial cells (Strain HeLa). Proc. Soc. exp. Biol. (N.Y.) 87, 480-485.

SCOTT, W.J., RITTER, E.J. & WILSON, J.G. (1975) Studies on induction of polydactyly in rats with cytosine arabinoside. Dev. Biol. 45, 103-112.

SCOTT, W.J., RITTER, E.J. & WILSON, J.G. (1977) Delayed appearance of ectodermal cell death as a mechanism of polydactyly induction. J. Embryol. exp. Morph. 42, 93-104.

SHELLSWELL, G.B. & WOLPERT, L. (1977) The pattern of muscle and tendon development in the chick wing. In: D.A. Ede, J.R. Hinchliffe, and M. Balls (Eds.), Vertebrate Limb and Somite Morphogenesis, Cambridge University Press, Cambridge, pp. 71-87.

SLACK, J.M.W. (1976) Determination of polarity in the amphibian limb. Nature, Lond. 261, 44-46.

SLACK, J.M.W. (1977) Control of anteroposterior pattern in the axolotl forelimb by a smoothly graded signal. J. Embryol. exp. Morph. 39, 169-182.

SMITH, J.C. (1979) Evidence for a positional memory in the chick limb bud. J. Embryol. exp. Morph. 52, 105-113.

SMITH, J.C. TICKLE, C. & WOLPERT, L. (1978) Attenuation of positional signalling in the chick limb by high doses of γ-radiation. Nature, Lond. 272, 612-613.

STOKER, M. & MACPHERSON, I. (1964) Syrian hamster fibroblast cell line BHK21 and its derivatives. Nature, Lond. 203, 1355-1357.

SUMMERBELL, D. (1974) Interaction between the proximo-distal and antero-posterior co-ordinates of positional value during specification of positional information in the early development of the chick limb-bud. J. Embryol. exp. Morph. 32, 227-237.

SUMMERBELL, D. (1979) The zone of polarizing activity: evidence for a role in normal chick limb. J. Embryol. exp. Morph. 50, 217-233.

SUMMERBELL, D., LEWIS, J.H. & WOLPERT, L. (1973) Positional information in chick limb morphogenesis. Nature, Lond. 244, 492-496.

SUMMERBELL, D. & TICKLE, C. (1977) Pattern formation along the antero-posterior axis of the chick limb bud. In: D.A. Ede, J.R. Hinchliffe, and M. Balls (Eds.) Vertebrate Limb and Somite Morphogenesis, Cambridge University Press, Cambridge, pp. 41-55.

TICKLE, C., SHELLSWELL, G., CRAWLEY, A. & WOLPERT, L. (1976) Positional signalling by mouse limb polarising region in the chick wing bud. Nature, Lond. 259, 396-397.

TICKLE, C., SUMMERBELL, D. & WOLPERT, L. (1975) Positional signalling and specification of digits in chick limb morphogenesis. Nature, Lond. 254, 199-202.

WHITTINGHAM, D.G., LEIBO, S.P. & MAZUR, P. (1972) Survival of mouse embryos frozen to -196°C and -269°C. Science, N.Y. 178, 411-414.

WOLPERT, L. (1971) Positional information and pattern formation. Curr. Top. devl. Biol. 6, 183-224.

WOLPERT, L. (1976) Mechanisms of limb development and malformation. Brit. Med. Bull. 32, 65-70.

WOLPERT, L., LEWIS, J.H. & SUMMERBELL, D. (1975) Morphogenesis of the vertebrate limb. In: R. Porter and J. Rivers (Eds.) Cell Patterning, Ciba Foundation Symposium 29, Elsevier, Amsterdam, pp. 95-130.

ZWILLING, E. (1956) Interaction between limb bud ectoderm and mesoderm in the chick embryo. I. Axis establishment. J. exp. Zool. 132, 157-172.

ZWILLING, E. & HANSBOROUGH, L.A. (1956) Interaction between limb bud ectoderm and mesoderm in the chick embryo. III. Experiments with polydactylous limbs. J. exp. Zool. 132, 219-239.

© 1980 Elsevier/North-Holland Biomedical Press
Development in Mammals, Vol. 4, M.H. Johnson, editor

CELLULAR EVENTS IN THE EARLY DEVELOPMENT
OF SKELETAL MUSCLES

G. B. Shellswell

A.R.C. Meat Research Institute,
Langford, Bristol BS18 7DY, U.K.

1.   The origin of limb muscle cells
   1.1   The role of the somites
   1.2   Evidence from avian embryos

2.   The formation of muscle pattern
   2.1   Dorsal and ventral muscle masses
   2.2   Possible mechanisms for the division of the
      muscle masses

3.   Defects in muscle development
   3.1   Muscular dysgenesis
   3.2   Muscular dystrophy
   3.3   Muscle pattern in syndactylism and
      oligosyndactylism - muscle-tendon autonomy

4.   A scheme for muscle development in the limb

5.   Conclusions

INTRODUCTION

Skeletal muscle has been a favourite tissue for the study
of cytodifferentiation because there is a precise series of
observable events which lead from myoblast to fully differentiated muscle fibre. Detailed investigations into the cytology
and biochemistry of muscle cytodifferentiation are possible
because this process will occur quite readily *in vitro* and thus
becomes readily accessible to experimental investigation. Indeed,
studies of this nature would be very difficult *in vivo* because of
the asynchrony of normal muscle development. Thus we are gaining
more and more information on how certain cells achieve those
properties characteristic of muscle differentiation.

Cytodifferentiation is only one facet of muscle development and there are other questions equally relevant to how skeletal muscle becomes a functional tissue: from where do muscle cells originate in the embryo? How do the differentiating myotubes become organised to produce separate muscles? These questions aroused the interest of embryologists in the early part of this century when embryology proved a useful tool in studies of the comparative morphology of different species. In recent years, there has been renewed interest in these issues and the chick embryo has featured prominently, due to its ease of access for experimental manipulation. The purpose of this review is to see how the results of these studies on muscle pattern formation and morphogenesis relate to the similar processes in mammalian myogenesis. Most of the work to be described involves limb muscles but there is good reason to think that limb development in the chick and mammal follow similar rules, as discussed by Tickle in this volume.

The approach followed will be to consider the two important questions of the origin of skeletal muscle cells and their arrangement to form separate muscles. There will be some reference to certain diseases and disorders of muscle to see how they might improve our understanding of normal muscle development.

1. THE ORIGIN OF LIMB MUSCLE CELLS

1.1 THE ROLE OF THE SOMITES

There has long been controversy in the literature over the origin of limb muscle cells. Holtzer et al (1957) showed that the somites can give rise to muscle fibres but there has been disagreement over whether cells from the somites gave rise to muscle fibres in the vertebrate limb. Earlier histological studies on reptiles (reviewed in Raynaud, 1977) revealed long ventral processes of cells linking the myotomes of the somites and the limb rudiment and it was generally held that cells in these processes would ultimately form the limb musculature. Similar, if slightly less distinct, columns of cells have been described in mouse embryos (Houben, 1976; Milaire, 1976) but the destination of these cells in later development was unknown.

The situation in human embryos has been equally confused. In his study on the developing human arm, Lewis (1901) cited eighteen studies concerned with the relationship between myotomes and limb buds before concluding that no distinct myotome buds enter the limbs of humans and higher vertebrates generally, although he could not discount the possibility that cells of myotomal origin enter the limb singly or in small groups. Corliss (1976) in his revision of Patten's classical text book on human embryology suggested a compromise where the proximal limb muscles are of somitic origin while distal muscles formed from condensations of somatopleural mesenchyme.

1.2 EVIDENCE FROM AVIAN EMBRYOS

A clearer understanding of the role of these somitically derived cells in limb development has come from the use of the chick-quail chimaeras, first described by Le Douarin and Barq (1969), in which markers are available which are particularly useful in the study of migrating cells. Implantation of somitic mesoderm from early stage quail embryos into similar positions in host chicken embryos of similar stages (usually about two days of incubation), followed by examination of cell distribution after a week's further development, has revealed that somitically derived (quail) cells were found in body muscles and that these cells had followed a myogenic rather than connective tissue type of differentiation (Chevallier, 1979; Christ et al, 1976, 1978). Experiments of this type have been used to map the origin of the cells in a number of body muscles, and it is now clear that somitic cells play the major role in providing cells which will later become muscle.

Turning to the wing, both groups of workers agree that cells derived from brachial somites are only found in limb muscle fibres, while cells of the other limb structures, cartilage elements, tendons and connective tissues including those of the muscle itself, are of somatopleural origin (see Figure 1 for the spatial relationship of these two types of mesoderm).

It is not clear whether all the skeletal muscle cells are somitically derived. Chevallier et al (1977, 1978) have reported muscle fibres containing both chick and quail nuclei in the limbs which developed after the reverse type of exchange (chick somites

140

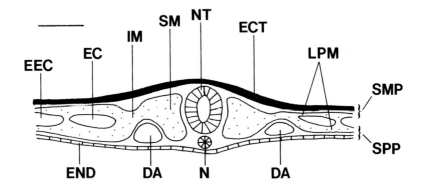

Figure 1: Diagrams to illustrate the spatial arrangement of trunk mesoderm
components at various stages in the development of the chick embryo.
(a: above) Stage 11, just less than 2 days incubation; (b: top opposite)
Stage 12+, 2 days incubation; (c: lower opposite) Stage 18, 3 days incubation.
Abbreviations: DA - dorsal aorta; DT - dermatome; EC - embryonic coelom;
ECT - ectoderm; EEC - extra-embryonic coelom; END - endoderm; IM - intermediate
mesoderm; L - limb; LPM - lateral plate mesoderm; MT - myotome; N - notochord;
NC - neural crest; NT - neural tube; SG - spinal ganglion; SM - somitic mesoderm; SMP - somatopleure; SPP - splanchnopleure; ST - sclerotome. Scale bar
represents 100μm in all cases.

into quail hosts). They also noted some muscle fibres in the
chick wing after the removal of brachial somites. However, Christ
et al (1979) suggested that these findings might have been caused
by regeneration of host somites. They reported that when quail
somatopleural mesoderm from the region that is going to form the
limb was grafted in place of the equivalent region in the chick,
then the resulting limb showed no indication that graft cells form
myotubes. This result indicates that somatopleural mesoderm does
not give rise to muscle. While it is true that the chick-quail
system suffers from the fact that some quail cells might appear as

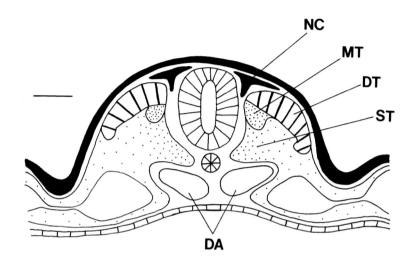

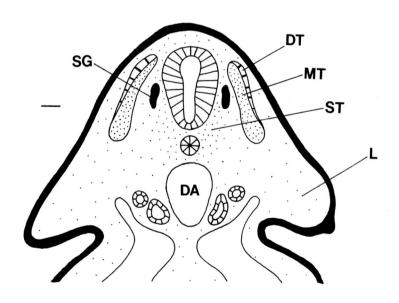

chick because the histological sections have missed the nucleolar
marker, Christ *et al* (1979) could find no evidence of quail muscle
cells in 50 grafts of this type so it is most unlikely that these
results were caused by sampling difficulties.

Christ *et al* (1979) also tested the ability of somatopleural
cells to undergo muscular differentiation in the absence of
somites. They found that isolated somatopleural mesoderm and
adjacent ectoderm grafted to the coelomic cavity of chick embryos
gave rise to fairly normal limbs containing well-developed
skeletal elements and tendons but muscular tissue was completely
absent. Unfortunately, these authors do not give information on
how many explants were performed or what stages of donor were
employed (myogenic cells start to migrate from wing-level somites
at about stage 14 Jacob *et al*, 1978; stages according to
Hamburger and Hamilton, 1951). There is also the possibility that
there had been muscle in the limbs which had degenerated due to
the absence of nerves by the time the limbs were examined, some
10 or 11 days after grafting. This information is essential now
that McLachlan and Hornburch (1979) have performed similar experiments using somatopleural mesoderm from the wing level and found
that grafts from as early as stage 10 (11 pairs of somites) could
give rise to muscle. They concluded that in the absence of the
somitic cells which normally give rise to the wing musculature,
somatopleural cells can perform this function.

The situation thus requires further clarification, with due
attention paid to both the sampling difficulties mentioned above
and the difference in timing of development between chick (21 days
to hatching) and quail (16 days) which might go some way to
explaining some of the discrepancies. It would also be interesting
to investigate the origin of muscle satellite cells. If these
cells, which are known to take part in regeneration of muscle
fibres, were shown to be of somatopleural origin, this could
account for the myogenic potential of some somatopleural cells in
unusual situations (such as in the absence of somitically derived
cells).

However, two points from these experiments on avian embryos
are beyond doubt: firstly, that somitic cells do invade the prospective limb mesoderm and secondly, that once in the limb,

somitic cells do not differentiate in any way except to form
muscle fibres.

These experiments provide useful pointers into the situation
in mammalian embryos and reinforce a suggestion of Harrison (1963)
that migration of somitic cells into limb buds of the human
embryo might occur when these cells are relatively undifferentiated
as muscle. It would be interesting to perform cross-species
grafting of mouse somites into chick embryos and then follow any
migration and development of the mouse cells either by immunological, surface-dependent techniques or by nuclear staining
(Moscona, 1957).

2. THE FORMATION OF MUSCLE PATTERN

2.1 DORSAL AND VENTRAL MUSCLE MASSES

The first morphological indication of myogenesis in the
tetrapod limb is the formation of muscle masses that are dorsal
and ventral to the chondrogenic core. Romer (1922) suggested that
the limb muscles of all tetrapods evolved from the two opposed
muscle masses of the fish fin. This theory has been widely supported by work describing two limb muscle masses in the early
development of the leg musculature of the chick (Romer, 1927;
Wortham, 1948) and lizard *Lacerta* (Romer, 1942) and in the
shoulder musculature of the newt *Necturus* (Chen, 1935), the lizard
Lacerta (Romer, 1944), the turtle *Chrysemys* (Walker, 1947), the
opossum *Didelphys* (Cheng, 1955) and the chick (Sullivan, 1962).

Repeated divisions and subdivisions of these muscle masses
produce the pattern of discrete limb muscles characteristic of
each of the animals in the above studies. Sullivan (1962) states
that the overall sequence of divisions and cleavages is similar
in all those tetrapods so far studied, and his conclusions were
verified in the mouse and mole by Milaire (1963) - see Figure 2.
It is for this reason that comparative embryology has revealed a
number of muscle homologues between different species which would
be missed in a study of the adult muscle pattern.

The similarities in the process of muscle mass division in
different species argue strongly for a common underlying mechanism.
In view of the central role of this process in the formation of the
often complex pattern of muscles necessary for limb movement, it

144

is surprising that few recent studies have examined it in detail.
The renewed interest in muscle development following from the
discovery of the somitic origin of limb muscle cells can shed new
light on possible mechanisms. Although most of the relevant
experiments have been done on the chick limb, from the similarities
discussed above, mammals may well behave comparably.

2.2   POSSIBLE MECHANISMS FOR THE
DIVISION OF THE MUSCLE MASSES

2.2.1 Tension in the growing limb

Carey (1921) and Sullivan (1962) suggested that tension
exerted by the longitudinal growth of the limb stimulates the
muscle masses to separate in the chick wing. There is no doubt
that there is a considerable lengthening of the wing while the
muscle masses divide – Summerbell (1976) describes a doubling of
the length of the forearm region over this period. Although many
authors have shown that tension is important for growth and
repair after injury (reviewed by Goldspink, 1974), there are few
studies implicating it in early events in myogenesis. Indeed,
various authors have demonstrated well formed muscle groups in
abnormally short limbs produced either by grafting limb buds to
the chorioallantoic membrane (Hunt, 1932; Bradley, 1970) or by
treatment in ovo with a teratogenic agent such as 6-aminonicoti-
namide, as used by McLachlan (1980). Thus tension may be excluded
as a possible cause of the division of the muscle masses.

2.2.2 Action of nerves

Hunt (1932) showed that although muscle would differentiate
in chick limb buds grafted to the chorioallantoic membrane, it was
only divided into distinct bundles with well developed associated
fibrous tissue if the grafts included an abundant nerve supply.
Nerves might cause the muscle masses to divide by direct mechanical
action, simply migrating into and pushing the muscle masses apart
(rather like cheesewire) as suggested by Wortham (1948) for the

Figure 2: Views of whole mouse embryos of various periods of gestation, with
sections through the mid-forearm regions. (a) and (b) opposite, 12 days;
(c) and (d) overleaf, 14 days; (e) and (f) overleaf, 17 days. Scale bars
represent: 1mm in a, c and e; 200µm in b, d and f. Note the progressive
divisions of the dorsal and ventral muscle masses (DMM, VMM).

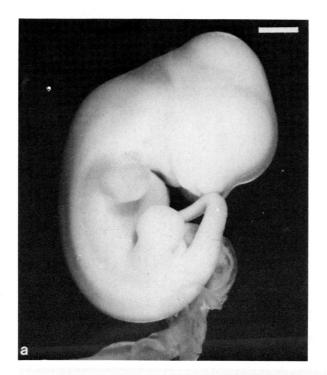

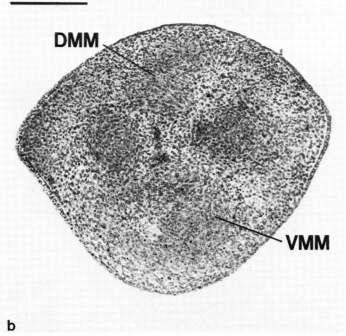

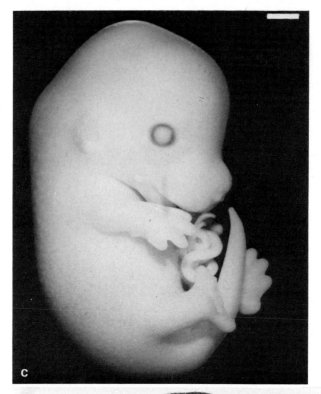

c

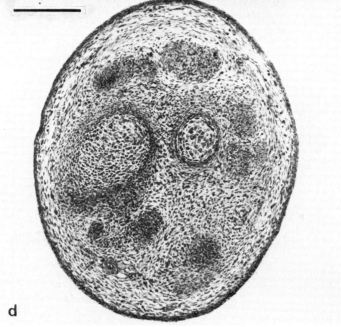

d

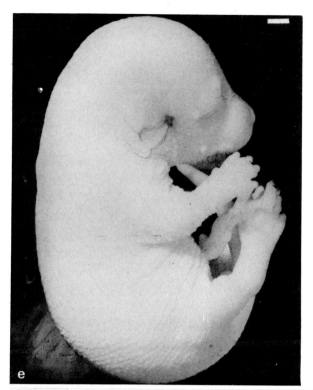

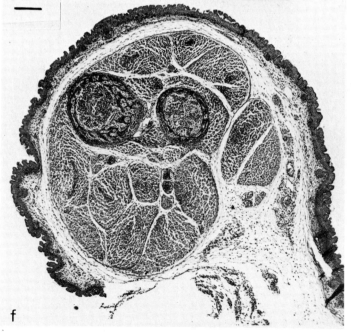

chick leg. Another possibility follows from the work of
Landmesser and Morris (1975) who found that they could cause contraction of different parts of the individual muscle masses by
differential nerve stimulation. If spontaneous stimulation of
this type were to occur in normal limb development, it could produce shearing movements within the muscle masses leading to
physical separation.

Shellswell (1977a) used low doses of the nicotinamide
analogue 3-acetylpyridine to remove peripheral nerves in the chick
wing before the muscle masses had completed their first division.
Treated embryos displayed a normal muscle pattern with frequent
elongated striated myofibrils. Furthermore, the excision of the
neural tube from the lumbar regions of the chick embryo at early
stages (2 days of incubation, before nerve outgrowth occurs at
this level) produces nerveless legs containing muscle masses which
nevertheless undergo the normal sequence of divisions, although
the myogenic cells degenerate at later stages (Jaros, personal
communication). Similar findings are reported by Grim and Carlson
(1979) who showed that normal patterns of muscle subdivisions can
occur in both embryonic and regenerating amphibian limbs in the
absence of nerves.

Thus while nerves appear necessary for the continued maintenance and appropriate cytodifferentiation (slow versus fast) of
muscle, as reviewed in Vrbova *et al* (1978), they do not play an
active role in the division of the limb muscle masses.

2.2.3 *Action of differentiating myoblasts*

A simply way to produce gaps in the muscle condensation would
be to promote localised regions of cell death. McMurrich (1923)
recognised cell degeneration as an important feature of the
morphogenesis of muscle and described how the decay of all or parts
of muscle segments often preceded the development of aponeuroses
of connective tissue. There is good evidence that cell death plays
a very important part in the morphogenesis of other tissues
(Saunders, 1966). Webb (1972, 1977) confirmed that cell death does
occur in the myogenic areas of the human fetus, but he found no
evidence for localised cell mortality at the presumptive gap
regions. Similar findings came from electron microscope studies of
the dividing muscle masses in the chick wing (Shellswell, 1977b).

The divisions of the muscle masses in the chick wing coincide with the phase of myoblast fusion and formation of myotubes and it may be that it is not necessary to stop (by cell death) this differentiation of myoblasts in the presumptive gap areas. A slowing down of myoblast differentiation in these areas would produce the desired cleavages as the myoblasts ran out of neighbours with which to fuse. If this were true, it would eventually lead to only a few myotubes in the gap areas as the muscle segments separate, and this result was found in the chick wing (Shellswell, 1977b). The effects of relatively small differences in the rates of differentiation would be magnified as development proceeded, thus producing the later subdivisions. A modulation of myoblast differentiation in this way could be achieved either directly, under the control of the system of positional information within the limb (Shellswell & Wolpert, 1977) or indirectly by way of changes in the extracellular environment of the myoblasts in the gap region. There is at present little evidence to support this hypothesis but it could bear further investigation.

2.2.4 *Action of non-muscle cells*

In their experiments on the somitic origin of chick limb muscle cells, Chevallier *et al* (1977) found that when somites from any cephalo-caudal level were heterotopically grafted to the prospective wing level, a normal pattern of wing muscles was always formed. This result shows that particular myogenic cells are not destined for particular muscles at the time they leave the somites. Wolpert (1978) has suggested that the pattern of muscles is determined by connective tissue cells. It may well be that it is *only* these somatopleurally-derived cells which respond to the postulated system of positional information within the limb as described by Shellswell and Wolpert (1977) for the chick.

Further evidence for this possible morphogenetic role of connective tissue cells comes from recent work by Christ *et al* (1979). These authors produced muscle-less limbs by grafting parts of quail somatopleural mesoderm and adjacent ectoderm into the coelomic cavity of host chick embryos. These limbs had well developed skeletal elements but no muscle. The interesting finding was that the mesenchyme of the muscle-forming zone was organised into definite lobes analogous to muscle blocks. However, there was no clear indication of the presence of the layers of connective tissue characteristic of muscle (the epi-, peri- and endo-mysium) and thus it may well be that the continued differentiation and development of muscle connective tissue is dependent on the presence of myogenic cells.

How might the connective tissue cells organise the pattern of muscles? Shellswell *et al* (1980) have investigated the involvement of one of the extracellular materials, collagen, in the division of the muscle masses. They conclude that collagen does not play an obvious mechanical role (such as a "wrapping up" of the muscle blocks) in separating the muscles. However, the amounts of collagen involved may have been too small for detection by the immunofluorescence methods of these authors, and there may also be some involvement of other extracellular materials (glycosaminoglycans and fibronectin) which play significant roles in other morphogenetic phenomena (see Vaheri *et al*, 1978; Morriss & Solursh, 1978; Saxen *et al*, this volume).

Wolpert (1978) has suggested that the formation and later subdivision of the dorsal and ventral muscle masses may be caused by a sorting-out process. He envisages sequential changes in the affinity of connective tissue cells for myogenic cells which eventually result in the formation of separate muscles. This type of process has been shown to underline the movement and patterning of mesenchymal cells in the sea urchin embryo (Gustafson & Wolpert, 1967). Such changes of affinity could operate by direct cell contact or via some component of the extracellular matrix produced by connective tissue, as discussed above.

In conclusion, although there is circumstantial evidence that the development of limb muscle pattern depends on some activity of connective tissue cells, a clearer understanding of how this might occur awaits further investigation.

3. DEFECTS IN MUSCLE DEVELOPMENT

Now that we are gaining more insight into the early development of normal muscle, what light does this throw on muscular disorders? Do the primary causes of these defects lie in these early stages? Of particular interest are the hereditary diseases, muscular dysgenesis and muscular dystrophy.

3.1 MUSCULAR DYSGENESIS

Pai (1965) and Platzer (1979) have investigated the embryology of murine muscular dysgenesis, a condition which results from an autosomal recessive mutation, lethal in the homozygous form *(mdg/mdg)*. Although both authors concentrate on the ultra-structural defects, there is evidence that the muscle mass is smaller and less divided in mutant limbs. There is a striking similarity here with the crooked neck dwarf mutant *(cn/cn)* in the chick, studied by Wick and Allenspach (1978). These latter authors reported the absence of epi- and peri-mysium in limbs of mutant embryos after 12 days of incubation, together with a lack of collagen and extracellular matrix in the subcutaneous connective tissue, although fibroblast-like cells were clearly present. They also reported diffusely organised tendons, and poor muscle-tendon attachments similar to those described in murine muscular dysgenesis by Banker (1977).

These observations suggest that muscular dysgenesis may be caused by a primary defect in the connective tissue which prevents the normal division of the muscle masses, as discussed above in Section 2.2.4. Unfortunately, Wick and Allenspach (1978) did not look at earlier embryos in which these changes should be more obvious. The other ultrastructural features of dysgenic muscle (retraction clots, abnormalities in nuclear shape and position, dilation of sarcoplasmic reticulum and disoriented myofilaments) could well be caused by defects in myotube differentiation, perhaps in response to some change in the relationship of myotubes to surrounding connective tissue. Possible candidates are a lack of collagen's normal stimulatory effect on myogenesis (noted *in vitro* by Hauschka & Konigsberg, 1966) or simple lack of tension, as suggested by Platzer (1977).

3.2 MUSCULAR DYSTROPHY

Platzer (1979) found that limb muscles from dystrophic mouse fetuses are morphologically normal and that ultrastructural alterations (variability in fibre size, coagulation necrosis,

retraction clots, fibre fragmentation and partial regeneration)
do not become clearly evident until birth or just after (6-9 days
post-partum). Webb (1974) has suggested that cell death, which
is a frequent feature of normal embryonic human myogenesis, is in
some way delayed or even absent in dystrophic embryos. He
suggests that this could lead to a higher incidence of "defective"
muscle fibres which normally die and disappear from the myogenic
population.

Thus it seems that dystrophy can be viewed as a defect
manifesting itself late in myogenic differentiation but there is
evidence that it is determined earlier, well before any visible
signs of cytodifferentiation, and once again this evidence comes
from work with the chick. Linkhart *et al* (1976) transplanted
limb buds between normal and dystrophic chick embryos at stage 19.
They found that cytodifferentiation of the muscle in the limbs
derived from these buds depended on their own origin and was
independent of later influence by the invading limb nerves of the
host type. Rathbone *et al* (1975) pushed the investigation back
to stage 13 and transplanted lengths of neural tube from
genetically dystrophic to normal embryos. They demonstrated an
induction, in genetically normal limb muscles, of abnormally
high thymidine kinase activity, the earliest enzyme change
characteristic of dystrophy in the chicken.

When the results of these two series of experiments are
taken with the more recent information on the somitic origin of
limb muscle cells in the chick, it is possible to suggest a site
for the primary action of the dystrophy mutation in a neural
tube - somite interaction. In dystrophic embryos, this interaction may occur just before the myogenic cells leave the somites
and results in the cells adopting a dystrophic type of differentiation. There is a need for somite transplantation experiments
(of the type described above in Section 1.2) between normal and
dystrophic embryos to further investigate this possibility and to
assess whether the results of Rathbone *et al* (1975) could have
been caused by contamination of neural tube grafts by some
somitic cells. Such experiments would also help to clarify the
precise timing of this proposed interaction. The question of
timing is important because the removal of the neural tube before

the somites become segregated in normal embryos does not affect
the early phases of myogenic differentiation such as myoblast
fusion and production of myofibrils (Jaros, personal communication).
Therefore, the "inductive" effect of dystrophic neural tube on
somites may be an additional step, merely modulating myogenic
differentiation which has been specified previously in somitic
cells by some as yet unknown mechanism.

In any event, somite transplantation may help to resolve the
long-running controversy over whether dystrophy is primarily
neurogenic or myogenic in origin (although it should be remembered
that there are differences between all animal models and human
dystrophies).

3.3 MUSCLE PATTERN IN SYNDACTYLISM AND OLIGOSYNDACTYLISM - MUSCLE-TENDON AUTONOMY

Milaire (1963) noted an autonomy between muscle and tendon
development in the mouse and mole where the formation of muscle-
tendon connections was secondary to the initial development of
both tissues. A similar degree of developmental autonomy of
muscle from tendon and cartilage has been described by Shellswell
and Wolpert (1977) and Kieny and Chevallier (1979) for early
stages in the chick wing, where distal tendons can develop in the
absence of their proximal muscles and *vice versa*. The latter
authors further conclude that the subsequent development and
maintenance of tendons nevertheless depends on some attachment to
a muscle block.

This autonomy is also illustrated in the syndactylous *(sm/sm)*
mouse, where Kadam (1962) and Milaire (1962, 1967) have described
deviations, reductions and absences of tendons which accompany
the digital malformations, while the muscles corresponding to
these tendons remain unaffected. In oligosyndactyly *(Os/+)*, where
digital abnormalities are more variable, distal tendons are
affected to a greater degree than in syndactyly. There is also
limited involvement of proximal muscles, particularly on the post-
axial side, while the digital abnormalities usually affect the
preaxial side.

All these studies reinforce the suggestion of Kieny and
Chevallier (1979) that the somatopleurally-derived cells, which
are involved in muscular organogenesis, can be divided into two

distinct compartments, the connective tissue of the muscle
itself and cells which will form the distal tendons, and that the
choice between these two compartments is probably genetically
controlled.

4. A SCHEME FOR MUSCLE DEVELOPMENT IN THE LIMB

It might be useful to summarise what is now known about the
development of skeletal muscle and thus synthesise an outline plan
of how it becomes a functional tissue. It should be stressed
that this is still very much a general outline, with many areas
yet to be filled in detail.

With very few exceptions, such as some muscles in the head,
the cells which form the contractile fibres of skeletal muscle
are derived from somitic mesoderm. The specification of this
type of cytodifferentiation seems to occur before the somitic
mesoderm becomes segregated into discrete somites. The mechanism
of the specification is, as yet, unknown but the neural tube may
well have some role if it depends on some assignment of positional
value to somitic cells.

Once their cytodifferentiation has been specified, myogenic
cells may contribute to body muscles either immediately via the
myotome or, as in limb muscles, after migrating to the appropriate
final site. In both cases, it appears that the final regional-
isation of myogenic cells is not determined at the same time as
their cytodifferentiation but occurs at some later stage, perhaps
after they have left the somite.

In the limb, the formation of the pattern of muscles follows
the formation of dorsal and ventral muscle masses which then
divide and subdivide until the adult pattern of muscles is formed
in miniature. It appears that the somatopleurally derived connective tissue plays a major part in this process. However, it is
clear that the formation of the pattern of muscles is an auto-
nomous process in that it can occur in the absence of the usual
distal points of insertion (either direct onto cartilage or by
distal tendons). When viewed in terms of positional information,
as discussed in Shellswell and Wolpert (1977), this suggests that
somatopleurally derived cells (perhaps those of the muscle connective tissue itself) respond to positional values set up within

the limb and thus determine the pattern of muscles, although the precise nature of this mechanism is again unknown.

The cytodifferentiation of the myogenic cells (myoblast fusion, maturation of myotubes) commences when the muscle masses start to divide, but its normal continuation depends on the presence of nerves. Nerves also exert a late influence on the type of cytodifferentiation ultimately followed by the muscle cells forming either fast or slow muscle fibres. This process occurs at peri-natal stages and details differ between species.

5. CONCLUSIONS

In this chapter I have described how recent experiments have influenced thinking on how muscle, predominantly limb muscle, develops. Much of this work has involved the chick embryo because of its ease of access. Although there is good morphological evidence that mammalian limbs develop in very similar ways, it is to be hoped that there will soon be experimental studies into muscle development in mammals. There are three approaches which could be explored to good effect. Firstly, there is the possibility of using cross-species grafting between mammalian and chick embryos, as employed in studying the mammalian limb zone of polarising activity (Tickle *et al*, 1976; see also Tickle, this volume). Secondly, parts of the mammalian embryo could be placed in organ culture as described by Agnish (1976) for mouse fore-limb rudiments; ways must be found to overcome the poor response of muscle, as compared to cartilage, under these conditions. Finally, there is the possibility of culturing whole mammalian embryos. This technique can be useful over the early stages of limb development although such embryos have yet to be grown beyond digit formation (15 days in the rat). Nevertheless, Deuchar (1976) showed that it is possible to perform surgical manipulations on the limb buds of early (11.5 days) rat embryos in culture by operating through the amnion and yolk sac. Cockroft (1976) has shown further that marginally older (12 days) rat embryos will grow with yolk sac and amnion opened and this renders accessible the first stages in myotube organisation within the limb muscle masses.

Hopefully, the above experimental approaches can be used to investigate these early stages of myogenesis so that there can be some attempt to ratify for mammalian embryos the picture of muscle development which is becoming clearer from the study of other species.

ACKNOWLEDGEMENTS

I would like to thank Lewis Wolpert for inspiring and encouraging my interest in muscle development and John McLachlan for our many discussions on this topic. Thanks also to Allen Bailey, Grant Burleigh and Cheryll Tickle for their help and advice during the preparation of this chapter, and to Rod Mounsdon for his photographic expertise. I am grateful to the Medical Research Council for their support.

REFERENCES

AGNISH, N.D. (1976) Possible role of somites in developing mouse limbs, in vitro. Anat. Rec. 184, 340-341.

BANKER, B.Q. (1977) Muscular dysgenesis in the mouse (mdg3mdg). I. Ultra-structural study of skeletal and cardiac muscle. J. Neuropathol. Exp. Neurol. 36, 100-127.

BRADLEY, S.J. (1970) An analysis of self differentiation of chick limb buds in chorio-allantoic grafts. J. Anat. 107, 479-490.

CAREY, E.J. (1921) Studies in the dynamics of histogenesis. IV. Tensions of differential growth as a stimulus to myogenesis in the limb. Amer. J. Anat. 29, 93-115.

CHEN, H.K. (1935) Development of the pectoral limb of Necturus maculosus. Illinois Biol. Monogr. 14, 1-77.

CHENG, C.C. (1955) The development of the shoulder region of the opossum Didelphys virginiana with special reference to the musculature. J. Morph. 97, 415-471.

CHEVALLIER, A. (1979) Role of the somitic mesoderm in the development of the thorax in bird embryos. II. Origin of thoracic and appendicular musculature. J. Embryol. exp. Morph. 49, 73-88.

CHEVALLIER, A., KIENY, M. & MAUGER, A. (1977) Limb-somite relationship: origin of the limb musculature. J. Embryol. exp. Morph. 41, 245-258.

CHEVALLIER, A., KIENY, M. & MAUGER, A. (1978) Limb-somite relationship: effect of removal of somitic mesoderm on the wing musculature. J. Embryol. exp. Morph. 43, 263-278.

CHRIST, B., JACOB, H.J. & JACOB, M. (1976) Uber die Herkunft der Mn. pectorales major et minor. Experimentelle Untersuchungen an Wachtel - und Huhnerembryonen. Verh. anat. Ges. Jena 70, 1007-1011.

CHRIST, B., JACOB, H.J. & JABOB, M. (1978) An experimental study on the relative distribution of the somitic and somatic plate mesoderm to the abdominal wall of avian embryos. Experientia 34, 241-242.

CHRIST, B., JACOB, H.J. & JACOB, M. (1979) Differentiating abilities of avian somatopleural mesoderm. Experientia 35, 1376-1378.

COCKROFT, D.L. (1976) Comparison of in vitro and in vivo development of rat foetuses. Develop. Biol. 48, 163-172.

CORLISS, C.E. (1976) Patten's Human Embryology - Elements of Clinical Development. McGraw-Hill, New York.

DEUCHAR, E.M. (1976) Regeneration of amputated limb-buds in early rat embryos. J. Embryol. exp. Morph. 35, 345-354.

GOLDSPINK, G. (1974) Development of muscle. In: G. Goldspink (Ed.), Differentiation and growth of cells in vertebrate tissues. Chapman and Hall, London, pp. 69-99.

GRIM, M. & CARLSON, B.M. (1979) The formation of muscles in regenerating limbs of the newt after denervation of the blastema. J. Embryol. exp. Morph. 54, 99-111.

GUSTAFSON, T. & WOLPERT, L. (1967) Cellular movement and contact in sea urchin morpholgenesis. Exp. Cell. Res. 83, 325-336.

HAMBURGER, V. & HAMILTON, H.L. (1951) A series of normal stages in the development of the chick embryo. J. Morph. 88, 49-92.

HARRISON, R.G. (1963) Textbook of human embryology. 2nd edition. Blackwell, Oxford, U.K.

HAUSCHKA, S.D. & KONIGSBERG, I.R. (1966) The influence of collagen on the development of muscle colonies. Proc. Natl. Acad. Sci. U.S.A. 55, 119-126.

HOLTZER, H., MARSHALL, J. & FINK, H. (1957) An analysis of myogenesis by the use of fluorescent antimyosin. J. biophys. biochem. Cytol. 3, 705-723.

HOUBEN, J-J. G. (1976) Aspects ultrastructuraux de la migration de cellules somitiques dans les bourgeons de membres posterieurs de souris. Arch. Biol. (Bruxelles) 87, 345-365.

HUNT, E.A. (1932) The differentiation of chick limb buds in chorio-allantoic grafts with special reference to the muscles. J. exp. Zool. 249, 301-326.

JACOB, M., CHRIST, B. & JACOB, H.J. (1978) On the migration of myogenic stem cells into the prospective wing region of chick embryos. A scanning and transmission electron microscope study. Anat. Embryol. 153, 179-193.

KADAM, K.M. (1962) Genetical studies on the skeleton of the mouse. XXXI. The muscular anatomy of syndactylism and oligosyndactylism. Gen. Res. Camb. 3, 139-156.

KIENY, M. & CHEVALLIER, A. (1979) Autonomy of tendon development in the chick wing. J. Embryol. exp. Morph. 49, 153-165.

LANDMESSER, L. & MORRIS, D.G. (1975) The development of functional innervation in the hindlimb of the chick embryo. J. Physiol. Lond. 249, 301-326.

LE DOUARIN, N. & BARQ, G. (1969) Sur l'utilisation des cellules de la caille japonaise comme "marqquers biologiques" en embryologie experimentale. C.R. Hebd. Seances Acad. Sci., Ser. D. Sci. Nat. (Paris) 269, 1543-1546.

158

LEWIS, W.H. (1901) The development of the arm in man. Am. J. Anat. 1, 145-184.

LINKHART, T.A., YEE, G.W., NIEBERG, P.S. & WILSON, B.W. (1976) Myogenic defects in muscular dystrophy in the chicken. Devel. Biol. 48, 447-457.

McLACHLAN, J. (1980) The effect of 6-aminonicotinamide on limb development. J. Embryol. exp. Morph., in press.

McLACHLAN, J.C. & HORNBRUCH, A. (1979) Muscle forming potential of the non-somitic cells of the early avian limb bud. J. Embryol. exp. Morph 54, 209-209-217.

McMURRICH, J.P. (1923) The development of the Human Body, 7th edn. McGraw-Hill, New York.

MILAIRE, J. (1962) Detection histochimique des modifications des ebauches dans les membres en formation chez la souris oligosyndactyle. Bull. Acad. Roy. Belg. 48, 505-528.

MILAIRE, J. (1963) Etude morphologique et cytochimique du developpement des membres chez la souris et chez la taupe. Arch. Biol. (Liege) 74, 129-317.

MILAIRE, J. (1967) Histochemical observations on the developing foot of normal, oligosyndactylous (Os/+) and syndactylous (sm/sm) mouse embryos. Arch. Biol. (Liege) 78, 223-288.

MILAIRE, J. (1976) Contribution cellulaire des somites a la genese des bourgeons de membres posterieurs chez la souris. Arch. Biol. (Bruxelles) 87, 315-343.

MORRISS, G.M. & SOLURSH, M. (1978) Regional differences in mesenchymal cell morphology and glycosaminoglycans in early neural-fold stage rat embryos. J. Embryol. exp. Morph. 46, 37-52.

MOSCONA, A. (1957) The development in vitro of chimaeric aggregates of dis-sociated embryonic chick and mouse cells. Proc. Natl. Acad. Sci. U.S.A. 43, 184-194.

PAI, A. (1965) Developmental genetics of a lethal mutation, muscular dysgenesis (mdg) in the mouse. II. Developmental analysis. Devel. Biol. 11, 93-109.

PLATZER, A.C. (1977) Myogenesis in mutant muscle: a model system for the role of contraction in development. J. Cell Biol. 75, 321a.

PLATZER, A. (1979) Emrbyology of two murine muscle diseases: muscular dystrophy and muscular dysgenesis. Ann. N.Y. Acad. Sci. 317, 94-113.

RATHBONE, M.P., STEWART, P.A. & VETRANO, F. (1975) Dystrophic spinal cord transplants induce abnormal thymidine kinase activity in normal muscles. Science 189, 1106-1107.

RAYNAUD, A. (1977) Somites and early morphogenesis of reptile limbs. In: D.A. Ede, J.R. Hinchliffe and M. Balls (Eds.). Vertebrate limb and somite morphogenesis. Symp. Brit. Soc. Devel. Biol. Vol.3, Cambridge University Press, Cambridge, pp. 373-385.

ROMER, A.S. (1922) Locomotor group of certain primitive and mammal-like reptiles. Bull. Amer. Mus. Nat. Hist. 46, 517-606.

ROMER, A.S. (1927) The development of the thigh musculature of the chick. J. Morph. Physiol. 43, 347-385.

ROMER, A.S. (1942) The development of tetrapod limb musculature - the thigh of Lacerta. J. Morph. 71, 251-298.

ROMER, A.S. (1944) The development of tetrapod limb musculature - the shoulder region of Lacerta. J. Morph. 74, 1-41.

SAUNDERS, J.W. (1966) Death in embryonic systems. Science, N.Y. 154, 604-612.

SHELLSWELL, G.B. (1977a) The formation of discrete muscles from the chick wing dorsal and ventral muscle masses in the absence of nerves. J. Embryol. exp. Morph. 41, 269-277.

SHELLSWELL, G.B. (1977b) Studies on the development of the pattern of muscles and tendons in the chick wing. Ph.D. thesis, University of London.

SHELLSWELL, G.B. & WOLPERT, L. (1977) The pattern of muscle and tendon development in the chick wing. In: D.A. Ede, J.R. Hinchliffe and M. Balls (Eds.). Vertebrate limb and somite morphogenesis. Symp. Brit. Soc. Devel. Biol. Vol.3, Cambridge University Press, Cambridge, pp. 71-86.

SHELLSWELL, G.B., BAILEY, A.J., DUANCE, V.C. & RESTALL, D.J. (1980) The role of collagen in the formation of the pattern of muscles in the developing chick wing. J. Embryol. exp. Morph. in press.

SULLIVAN, G.E. (1962) Anatomy and embryology of the wing musculature of the domestic fowl (Gallus). Aust. J. Zool. 10, 458-518.

SUMMERBELL, D. (1976) A descriptive study of the rate of elongation and differentiation of the skeleton of the developing chick wing. J. Embryol. exp. Morph. 35, 241-260.

TICKLE, C., SHELLSWELL, G., CRAWLEY, A. & WOLPERT, L. (1976) Positional signalling by mouse limb polarising region in the chick wing bud. Nature (London). 259, 396-397.

VAHERI, A., RUOSLAHTI, E. & MOSHER, D.F. (1978) (Eds.). Fibroblast Surface Protein. Annals N.Y. Acad. Sci. 312, 1-456.

VRBOVA, G., GORDON, T. & JONES, R. (1978) Nerve-Muscle Interaction. Chapman and Hall, London.

WALKER, W.F. (1947) The development of the shoulder region of the turtle Chrysemys peta marginata with special reference to the primary musculature. J. Morph. 80, 193-249.

WEBB, J.N. (1972) The development of human skeletal muscle with particular reference to muscle cell death. J. Path. 106, 221-228.

WEBB, J.N. (1974) Muscular dystrophy and muscle cell death in normal foetal development. Nature 252, 233-234.

WEBB, J.N. (1977) Cell death in developing skeletal muscle: histochemistry and ultrastructure. J. Path. 123, 175-180.

WICK, R.A. & ALLENSPACH, A.L. (1978) Histological study of muscular hypoplasia in the crooked neck dwarf mutant (cn/cn) chick embryo. J. Morph. 158, 21-30.

WOLPERT, L. (1978) Pattern formation in biological development. Sci. Amer. 239, 124-137.

WORTHAM, R.A. (1948) The development of the muscles and tendons in the lower leg and foot of chick embryos. J. Morph. 83, 105-148.

© 1980 Elsevier/North-Holland Biomedical Press
Development in Mammals, Vol. 4, M.H. Johnson, editor

MECHANISMS OF MORPHOGENETIC CELL INTERACTIONS

Lauri Saxen, Peter Ekblom and Irma Thesleff

Department of Pathology,
University of Helsinki,
Finland

1. INTRODUCTION

1.1 DEFINITION

An organism, organ or any mature tissue consists of various tissue components and cell types arranged spatially into functional entities. During embryogenesis, these components develop spatially and temporally in a strictly synchronized manner. This suggests that each tissue and cell must be capable of communicating with the cells and tissues in close association with it. Cells release signals to be received by others, and each of the cells must also develop mechanisms for interpreting the signals in order to convert them into a response which has morphogenetic meaning. We call this dialogue between two closely-associated cell populations of different origin "morphogenetic or inductive cell interaction" and the future fate of these cell populations is determined by the interaction. This definition excludes a number of developmentally significant control mechanisms such as guided cell migrations, long-range stimulation of proliferation (growth factors) and the morphogenetic action of hormones. Interactions between similar cells, their mutual recognition, aggregation and structural organization are also excluded by this definition. The mode of response to inductive interactions by the target cell may include gene activation, increased or inhibited cell proliferation, altered surface properties or even by a developmentally significant lysis (morphogenetic cell death).

Therefore our definition conforms to that proposed by Grobstein (1956a) according to which induction takes place when *"two or more tissues of different history and properties become intimately associated and alteration of the developmental course of the interactants results"*. The significance of morphogenetic cell interactions and their universal distribution among higher animals has been established by experimental investigations undertaken since early in this century (Reviews: Spemann, 1938; Grobstein, 1956a, 1967; Saxen & Toivonen, 1962; Kratochwil, 1972, Saxen *et al*, 1976a; Wessells, 1977).

1.2 SIGNAL SUBSTANCES

Biological interactions involve a chain of events at the molecular level: the triggering of the synthesis and release of a signal substance, its transmission to the target and finally, a response of the target. As far as morphogenetic cell interactions are concerned, this chain of events has not been fully analyzed in any interactive system, and the signal substances operative *in vivo* are virtually unknown. In what follows, we shall examine a number of substances as candidates for this role, including

inorganic ions, morphogenetic proteins, neurotransmitters and
components of the cell surface, the cell periphery and the intercellular matrix.

1.3 TRANSMISSION OF SIGNALS

For as long as the actual chemical nature of the signal
substances is unknown, indirect information must be obtained
from observations on the localization of signals in the tissues as
well as on their mode of transmission. The dispute concerning
whether the signals involve diffusible substances or cell-mediated
interactions is an old one. The first experiments relating to
this problem date back to 1932 (Bautzmann et al.) and in 1953
Grobstein made the first successful transfilter experiments.
Later he presented a third type of possible transmission mechanism
for morphogenetic signals. In addition to the original hypothesis
of diffusible compounds (Bautzmann et al, 1932) and the postulate
of contact-mediated interactions (Spemann, 1901; Weiss, 1949),
Grobstein (1956a) also stressed the potential significance of components of the intercellular matrix.

These alternative transmission mechanisms have subsequently
been thoroughly examined in a number of model-systems. Basically,
two approaches have been practised: (i) physical or chemical
interference with the transmission of signals, and (ii) direct
(ultrastructural, histo- and immuno-chemical) observations on cell
relations during morphogenetic communication. Furthermore, cell-
free extracts, either free or bound to particles, have been
employed as well as labelled inductors (tissues, their fractions
and defined molecules). Referring to experiments like this, all
three transmission mechanisms originally suggested by Grobstein
(1956a) still appear plausible (Figure 1). We shall examine these
hypotheses in some selected model-systems of morphogenetic cell
interactions and also discuss the nature of the molecules
involved and their mode of action.

2. DIFFUSION OF SIGNAL SUBSTANCES

The long-range diffusion of compounds carrying morphogenetic
messages has long been the favourite model of many developmental
biologists, but there is not much evidence for compounds of this
kind in interactive events. Most conclusive evidence comes from

164

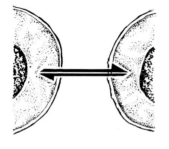 **DIFFUSION
OF SIGNAL
SUBSTANCES**

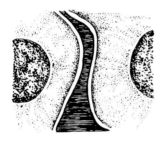 **MATRIX-
MEDIATED
INTERACTION**

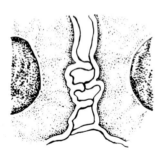 **CELL-
MEDIATED
INTERACTION**

Figure 1: Alternative modes of transmission of signal substances in morphogenetic cell interactions (After Grobstein,1956a and Saxen, 1972).

experiments on two successive interactions, the primary
neuralization of the gastrula ectoderm and the classic "secondary"
induction leading to the formation of the lens.

2.1 NEURAL INDUCTION

The determination of the presumptive neuroectoderm in
Amphibian gastrulae, with the invaginating chordamesoderm as a
neural inductor, is a classic model for induction.   Experiments

on the transmission of this inductive signal are complicated to some extent by the fact that the process has several steps. As the ectodermal cells become uniformly neuralized, a second control event leads to their transformation into the various regional elements of the CNS (Saxen et al, 1964; Nieuwkoop, 1973). It has proved difficult during this prolonged, multi-step process to relate direct ultrastructural observations on cell relations to the actual transfer of inductive signals. Intimate contacts between the ectodermal target cells and those of the inducing chordamesoderm have been described repeatedly (Eakin & Lehmann, 1957; Keller & Schoenwolf, 1977; Grunz & Staubach, 1979), but conclusive evidence for the transmission of morphogenetic messages via these contacts is still lacking.

Neural induction can be provoked by a variety of cell-free fractions and chemical compounds ranging from mono- and divalent cations (Masui, 1959; Barth & Barth, 1968, 1972; Johnen, 1970), proteins from chick embryo extracts (Tiedemann et al, 1969; Tiedemann, 1976, 1978), neurotransmitters (McMahon, 1974), cyclic nucleotides (Wahn et al, 1976) and heparan sulfate (Landstrom & Løvtrup, 1977). These experiments show that the normal living inductor tissue can be replaced by soluble factors, but do not show whether in the normal in vivo situation soluble factors are operating.

It is also possible to examine the contact prerequisites of the interactants in this system by separating the normal inductor tissue from its target cells by various membranes. By using a modification of Grobstein's (1956b) original filter technique, Saxen (1961, 1963) obtained neuralization of the presumptive neuroectoderm when it was separated from the living inductor tissue by a Millipore filter, 25µm thick and with an average pore size of 0.8µm (Figure 2). The electronmicroscopic examination of the filters failed to demonstrate cytoplasmic penetration into the pores (Nyholm et al, 1962). The problem was recently reconsidered when new filter membranes with better specifications and straight pores became available and the above observations were confirmed (Toivonen & Wartiovaara, 1976; Toivonen et al, 1976; Toivonen, 1979). Neural induction regularly occurred through all filters tested; pores down to 0.05µm still allowed

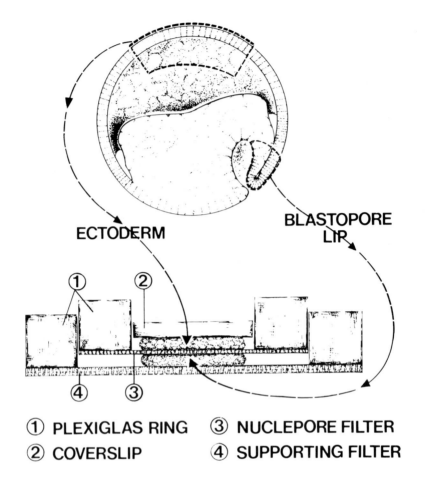

ECTODERM   BLASTOPORE
                                LIP

① PLEXIGLAS RING   ③ NUCLEPORE FILTER
② COVERSLIP   ④ SUPPORTING FILTER

Figure 2: Scheme of the transfilter experiments with tissues from Amphibian gastrulae (After Saxen, 1961).

the passage of signals, although no cytoplasmic processes were observed to have penetrated the filters (Figure 3).

The results for the subsequent transformation of the neuralized cells and the induction of mesodermal derivatives in the ectoderm are still slightly contradictory. Filters with relatively large pores (0.4 to 0.6µm) and a prolonged trans-filter contact allow both the transformation and the mesoderma-lization of the ectoderm (Minuth, 1978; Toivonen, 1979). Minuth has not detected cytoplasmic penetration of the filter, whereas Toivonen (1979) reports ectodermal cell processes in the filter,

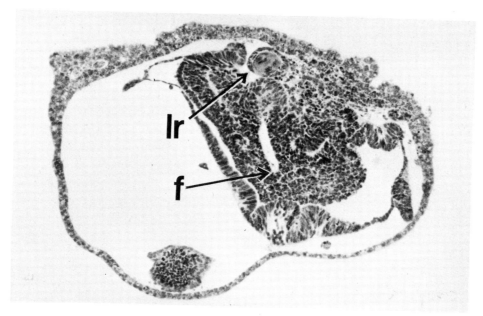

Figure 3: Neuralization of a gastrula presumptive ectoderm cultured trans-filter to the blastoporal lip. A Nuclepore filter with an average pore size of 0.05µm was employed. f, forebrain; lr, lens with an optic cup. (Courtesy of S. Toivonen).

making contact with the inductor. We are left with the still somewhat tentative conclusion that whereas the neuralization of the competent ectoderm is brought about by diffusible molecules released by the chordamesoderm, the subsequent transformation of the neuralized cells may require actual cell-to-cell contacts.

2.2 INDUCTION OF LENS

The vertebrate lens is induced from the epidermis by a series of morphogenetic cell interactions culminating in the terminal effect of the optic vesicle on the predetermined epidermal cells (Jacobson, 1958). This last step, highly specific *in vivo*, is "permissive" in nature and can be achieved without contact between inductor (optic cup) and epidermis by providing optimal culture conditions for the epidermis isolated before a contact is established (Mizuno, 1972; Karkinen-Jaaskelainen, 1978a). The conclusion that this permissive stimulus does not require an actual cell contact between the interactants is further supported by

168

several electron microscopic studies, failing to show cytoplasmic
continuity between the epidermis and the optic cup (Cohen, 1961;
Hunt, 1961; Weiss & Fitton-Jackson, 1961; Hendrix & Zwaan, 1974).

Recent evidence for the action of a diffusible compound has
been provided by Karkinen-Jaaskelainen (1978b). Unlike investigators before her (Muthukkaruppan, 1965), she exposed trunk rather
than predetermined head epidermis to the optic cup in a transfilter set-up (Figure 4). All filters used allowed the passage of
the inductive effect, and lentoid bodies with crystallin synthesis
formed from the trunk epidermis. Among the filters employed were
100µm thick Millipore filters and double Nucleopore filters with
an average pore size of 0.1µm - both known to prevent cytoplasmic
ingrowth. Finally, a dialyzer membrane was inserted between the

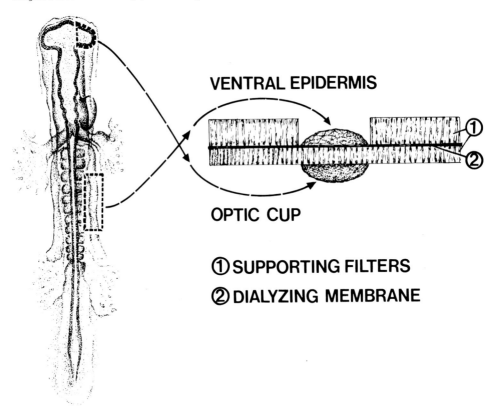

Figure 4: Scheme of the transfilter experiments on lens induction in chick
embryos (details: Karkinen-Jaaskelainen, 1978b).

tissues (allowing the penetration of molecules smaller than
12.000 daltons). In 21 out of 30 explants lentoid bodies
developed. It was therefore concluded that the signal substances
of lens induction could diffuse over distances up to 100μm, and
they had a molecular weight of less than 12.000 daltons.

2.3 EFFECTS OF DIFFUSIBLE INDUCERS

The molecular mechanisms of the inductive interactions
leading to the determination and segregation of the CNS and the
lens are not known. The ontogeny of the lens proteins strongly
suggests an effect at the level of gene expression (Clayton,
1977), but the mechanism by which an inductor might achieve these
changes is not clear. Clayton (1977) suggests that changes in
the cell surface might be important during the differentiation
and maturation of the lens fibre cells. She concludes with a
working hypothesis "that the membrane imposes some modulations
on the signals received and transmits these to effector systems
in the cells". This is basically what one would conclude from
the data available for neural induction as well.

Neuralization can be induced by a variety of experimental
treatments ranging from a physical shock and "sublethal" treatment (Holtfreter, 1955) to an exposure to protein fractions
(Tiedemann, 1976, 1978). The only common denominator for these
diverging, unspecific stimuli seems to be an effect on the plasma
membrane of the target cells. The precise mode of action of the
inorganic ions leading to neural induction is not known, but
Barth and Barth (1968, 1972) suggest that they may cause changes
in the permeability of the plasma membrane, and as a consequence
alter the ionic pool of the intracellular compartment. This, in
turn, could activate a second messenger (cyclic AMP, cyclic GMP),
leading thus to an ultimate effect of the genome and its
expression (McMahon, 1974). This hypothesis has not yet found
direct experimental evidence because the stimulation of the
gastrula ectoderm with cyclic nucleotides has yielded inconsistent
results (Wahn et al, 1976; Grunz & Tiedemann, 1977).

An interesting, though highly speculative idea was recently
proposed by Hay (1978). Referring to the results of Barth and
Barth (1968, 1972), she suggests that the chordamesoderm may
produce an "ionic filter matrix" under the neuroepithelium. She

points out that the microtubule-dependent elongation of the cells during neurulation is known to be highly sensitive to the intra-cellular ion concentration.

A special case of an apparently receptor-mediated induction of the gastrula ectoderm was recently reported by Kaska and Triplett (1980). When cultured *in vitro*, the explants of the chordamesoderm (somites) release two protein fractions. When ectodermal cells are cultured in this "conditioned" medium, they are induced to synthesise tyrosine oxidase. The proteins may be internally labelled and shown to bind to the ectodermal cells (but not to the entodermal cells), suggesting that the target cells contain specific receptors for the inductor molecules.

Tiedemann and Born (1978) have also suggested an effect of the neuralizing inductor on the plasma membrane. A fraction containing both the neuralizing factor(s) and the "vegetalizing" factor(s) (which induces meso- and ento-dermal derivatives) was covalently bound to sepharose beads and tested on competent gastrula ectoderm. The vegetalizing activity of the fraction was lost on binding, but restored after an enzymatic cleavage of the sepharose matrix. The neuralizing activity was unaffected by the binding procedure. The authors concluded that whereas the neural inductor may act upon the plasma membrane receptors without entering the cells, the vegetalizing factor apparently enters the cell to exert its action.

3. MATRIX-MEDIATED INTERACTIONS

In several organ systems the interacting tissues are separated by an extracellular matrix (ECM), and hence it has been reasonable to assume that interaction in such cases might be mediated by components of the ECM. Grobstein (1955a, 1967) was the first to stress the role of the ECM in morphogenetic tissue interactions, and his hypothesis has since gained support in a number of studies. Although some studies performed with the transfilter technique should be re-evaluated, because the filters designed for preventing cell contacts between the interacting tissues do not necessarily do so, there is evidence that differen-tiation is stimulated by interaction between the differentiating cells and the ECM (Slavkin & Greulich, 1975). Three systems of this kind will be discussed and the role of different ECM macromolecules in these interactions and in morphogenesis will be discussed.

3.1 INDUCTION OF SOMITE CHONDROGENESIS

Interaction of the notochord and the spinal cord with somites is required for the differentiation of the vertebral cartilage. The somite consists of the sclerotome and the dermomyotome, but only the sclerotomal mesenchyme chondrifies (Figure 5). This is because the sclerotomal cells alone migrate

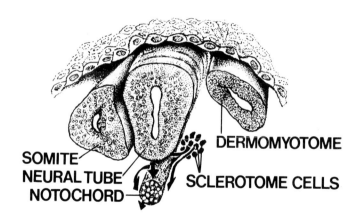

Figure 5: Scheme of somite differentiation in the chick embryo. Sclerotome cells migrate medially and undergo chondrogenic differentiation as they reach the extracellular matrix surrounding the notochord and the neural tube.

to the vicinity of the notochord and spinal cord, whose inductive influence has been well established in extirpation and transplantation experiments (Strudel, 1953; Holtzer & Detwiler, 1953; Watterson et al, 1954). The requirement for this tissue interaction has been demonstrated also   in vitro. Cultured with somitic mesenchyme, the notochord and spinal cord stimulate cartilage differentiation (Review: Lash & Vasan, 1977).

The notochord and the spinal cord secrete an ECM composed of proteoglycan and collagen. The glycosaminoglycans (GAGs) have been characterized predominantly as chondroitin 4- and 6-sulfate (Hay & Meier, 1974; Kosher & Lash, 1975) and the collagen as type II (Linsenmayer et al, 1973). Sclerotome cells migrate into this matrix just before their differentiation into cartilage (Ruggeri, 1972). In vitro studies have provided evidence for the involvement of the matrix components in the mechanism by which the notochord and spinal cord promote chondrogenesis. If the matrix is removed from the notochord or the spinal cord before culture with somitic mesenchyme, chondrogenesis is not stimulated until the ECM has reformed (Kosher & Lash, 1975).

The molecular profiles of the proteoglycans during somite chondrogenesis can be characterized by using molecular sieve chromatography (Lash & Vasan, 1978). It is possible to use the changes in proteoglycans as indices of differentiation and to analyze the effects of various ECM components on the profiles. During chondrogenic differentiation, the size of the proteoglycan molecule synthesized increases, and more and more proteoglycan aggregates are synthesized. The same changes were seen in vitro when somitic mesenchyme was cultured on a Nuclepore filter (Figure 6). After 14 days of culture the proteoglycan profile of the somitic mesenchyme was identical with that of the mature embryonic vertebral cartilage. This confirms that the somitic mesenchyme has an inherent chondrogenic bias. When the notochord was cultured with the mesenchyme, the rate of molecular changes in the proteoglycans was enhanced. Exogenous proteoglycans added to the culture medium slightly enhanced the change in the proteoglycan profile. Substrates of type I, II and III collagen stimulated changes in the proteoglycan profiles, type II collagen being the most effective. The chondrogenic stimulation was greatest when somitic mesenchyme was cultured on a matrix composed of type II collagen and proteoglycan aggregate (Figure 6). It was suggested that the correlation between type II collagen and the greatest stimulation has a developmental significance, as type II collagen is the major collagen type produced by the notochord, the in vivo inductor (Lash & Vasan, 1978).

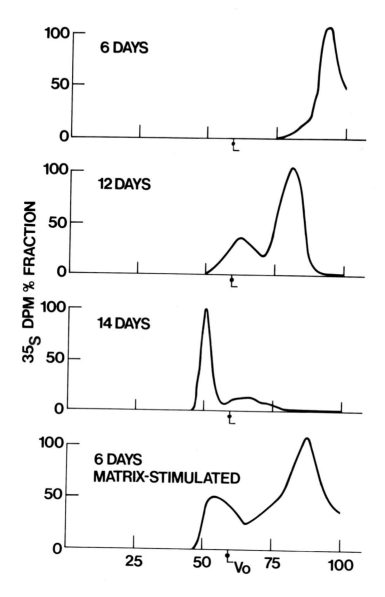

Figure 6: Molecular sieve chromatography of proteoglycans from extracts of
chick embryo somites, cultured either on a plain Nuclepore filter for 6, 12
and 14 days, or for 6 days on a filter covered with an artificial matrix containing proteoglycan aggregate and type II collagen. Proteoglycans to the left
of the void volume mark (V_0) represent proteoglycan aggregates, and the large
peaks to the right proteoglycan monomers. The profile of somites cultured for
14 days is identical with that of the mature embryonic cartilage. Comparison
with the lower profile reveals the stimulation effect of the artificial matrix
(After Lash & Vasan, 1978).

3.2 ENHANCEMENT OF SYNTHESIS OF THE
CORNEAL STROMA

The formation of the primary corneal stroma by the corneal
epithelium is another developmental system in which interaction
between cells and the ECM has been demonstrated. Corneal
epithelium differentiates from the head ectoderm overlying the
site of detachment of the lens vesicle (Figure 7). The epithelial
differentiation is characterized by transformation to a thick
secretory epithelium and the subsequent secretion of the primary
corneal stroma, consisting of an orthogonal array of collagen
fibrils embedded in GAGs (Hay & Revel, 1969; Coulombre &

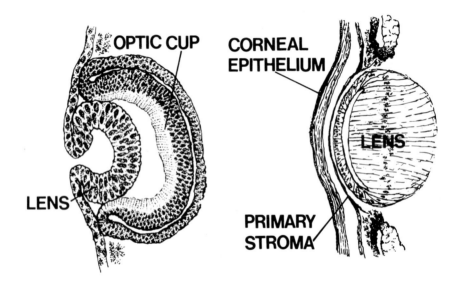

Figure 7: Scheme of the formation of the lens and cornea in the chick embryo.
The differentiation of the corneal epithelium is influenced by the lens and
the optic cup, and it is characterized by transformation to a thick secretory
epithelium with subsequent secretion of the primary corneal stroma.

Coulombre, 1971). It has been shown that the lens and optic cup
influence corneal differentiation via a close contact between the
interacting tissues (Lopashov & Stroeva, 1961). Because the lens,
the optic cup and the presumptive corneal epithelium are each
underlined by continuous basement membranes, it was suggested that

the induction of epithelial differentiation was mediated by the
ECM separating the interacting tissues (Hay & Revel, 1969).

The stimulatory effect of the ECM has been demonstrated in
numerous *in vitro* studies in which enzymatically removed corneal
epithelium has been grown on different substrata. The isolated
epithelium produced the primary corneal stroma if cultured on
living lens, on a killed lens capsule or on other collagenous
substrata (Dodson & Hay, 1971; Hay & Dodson, 1973). The production of stroma correlated with an increase in collagen and GAG
synthesis (Meier & Hay, 1974a). Collagen synthesis was not
affected by addition to the culture medium of GAGs (chondroitin
sulfate, heparin or heparan sulfate) but the synthesis by the
epithelium of chondroitin sulfate and heparan sulfate like compounds was enhanced 2-fold (Meier & Hay, 1974b). As any ECM containing collagen stimulated stroma production with its accompanying
increase in collagen and GAG synthesis, it was concluded that the
natural stimulation is a direct effect of the collagen of the ECM
(Hay & Meier, 1976). This conclusion was also supported by
experiments which showed that highly purified cartilage-type
collagen had the same stimulatory effect on stroma production as
the lens capsule.

The stimulatory effect of collagen on collagen synthesis by
the corneal epithelium requires contact between the epithelial
cells and the ECM, as was shown in experiments with corneal
epithelium grown transfilter from lens capsule or purified
cartilage collagen. The stimulatory effect was directly proportional to the size of the contact area between the epithelial cell
processes and the ECM (Figure 8; Hay & Meier, 1976). Furthermore,
this interaction seems to take place at the cell surface, as no
collagen was seen to enter the epithelial cells when lens capsules
were labelled radioactively.

3.3 *DIFFERENTIATION OF ODONTOBLASTS*

Odontoblasts differentiate from neural crest derived
mesenchymal cells in the bell-stage tooth germ. Their differentiation is characterized by the alignment of the cells under the
enamel epithelium, their polarization and subsequent secretion of
the organic matrix of dentin (Figure 9). This differentiation

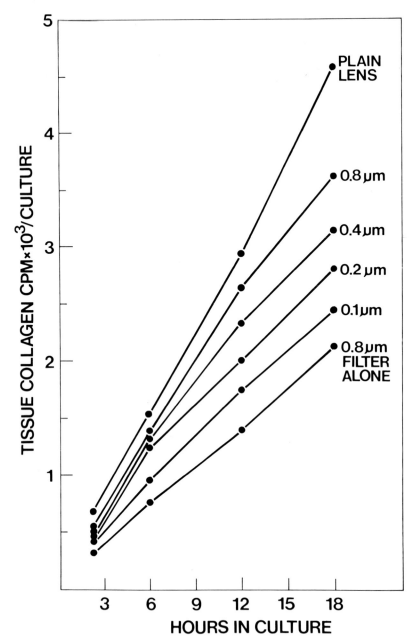

Figure 8: Stimulation by lens capsule of synthesis of collagen by corneal epithelium, grown directly on the lens capsule (top curve), on a plain Nuclepore filter (bottom curve), or separated from lens capsule by filters of different pore sizes. (After Meier & Hay, 1975).

process requires an epithelial-mesenchymal interaction (Gaunt & Miles, 1967), which may be mediated by the ECM at the epithelial basement membrane. *In vivo* the basal lamina is continuous under the enamel epithelium at the time of odontoblast differentiation, and it thus prevents actual cell-cell contacts between the interacting tissues (Kallenbach, 1971; Meyer *et al*, 1977). Transfilter experiments showed that diffusible matrix molecules from the epithelium did not induce odontoblast differentiation (Thesleff *et al*, 1977). In these studies, the epithelium and mesenchyme were separated enzymatically before the odontoblasts

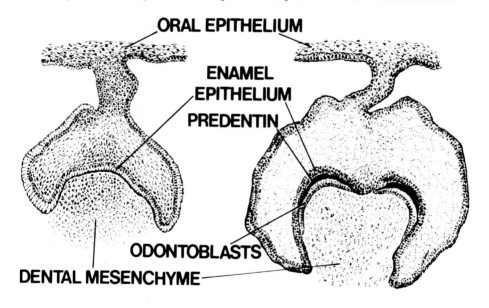

Figure 9: Scheme of the differentiation of odontoblasts in a tooth germ. The dental mesenchymal cells directly underlying the enamel epithelium become polarized odontoblasts and start the secretion of collagenous predentin.

had differentiated and were co-cultured separated by interposed Nuclepore filters of varying pore sizes. The differentiation of odontoblasts was seen as the alignment of the mesenchymal cells on the filter surface (Figure 10A), their polarization and subsequent secretion of collagenous predentin. The differentiation was prevented by filters with 0.1µm pore size, which also prevented the ingrowth of cell processes into the filter pores (Figure 10C,D). Ultrastructural examination of those filters

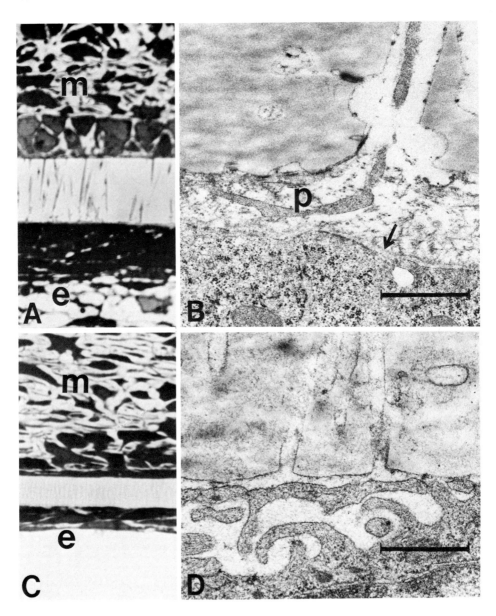

Figure 10: Micrographs of dental epithelium (e) and mesenchyme (m) cultured with interposed Nuclepore filters of 0.6μm (A,B) and 0.1μm diameter pores (C,D). The 0.6μm pore size filter allows the differentiation of odontoblasts, first seen as the alignment of the mesenchymal cells on the upper filter surface (A). This alignment is preceded by the restoration of the basement membrane (arrowed) and traversing of mesenchymal cell processes (p) through the filter (B). The 0.1μm pore size filter prevents both penetration of cell processes (D) and differentiation of the odontoblasts (C). (Scale bar = 1μm) (Thesleff *et al*, 1978).

allowing odontoblast differentiation revealed mesenchymal cell processes traversing the filter and terminating close to a continuous basal lamina that had been restored between the epithelial cells and the filters (Figure 10B; Thesleff *et al*, 1978). The results of these experiments suggest that the differentiation of odontoblasts is triggered by contacts between the mesenchymal cells and the basement membrane material.

3.4 *SUGGESTED ROLES OF VARIOUS ECM MACROMOLECULES*

The major classes of the ECM macromolecules that have been implicated in tissue interactions are collagen, glycosamino-glycans (GAG) and glycoproteins. Direct evidence for the role of an ECM molecule came first from studies by Konigsberg and Hauschka (1965) who demonstrated the stimulatory effect of collagen on muscle cell differentiation. Further evidence for collagen and GAGs as mediators of tissue interactions has since been derived from studies on somite chondrogenesis and corneal epithelial differentiation (above). These studies have clearly demonstrated that extracellular collagen or GAG of the same type that is produced by the inductor tissue *in vivo* (Hay & Meier, 1974) can stimulate *in vitro* the synthesis of similar molecules by the responding tissue. The actual mechanisms by which collagen and GAG stimulate differentiation are not known. In the case of somite chondrogenesis, it has been suggested that the macro-molecules act by lowering the level of intracellular cyclic AMP (Kosher, 1976). Collagen may stimulate corneal differentiation by providing a substrate for cell adhesion, which could result in changes at the inner surface of the plasma membrane (Hay, 1977).

In the case of the epithelio-mesenchymal interaction that leads to the differentiation of odontoblasts, the molecules of the ECM involved are not known. So far it has not been possible to stimulate odontoblast differentiation in isolated dental mesenchyme without living epithelial tissue (Thesleff, 1978). Our knowledge of the chemical composition of the basement membrane separating the interacting tissues has increased considerably during the last few years (Kefalides, 1978; Timpl *et al*, 1979a,b; Kanwar & Farquhar, 1979; Hassel *et al*, 1979). Collagenous and non-collagenous proteins as well as GAGS in the basement membranes

have been characterized, and subsequently localized, in various
basement membranes including those of the developing tooth germ
(Linder *et al*, 1975; Lesot *et al*, 1978; Thesleff *et al*, 1979;
Ekblom *et al*, 1980a; Foidart *et al*, 1980), and observed spatial
and temporal variations in the distribution of such molecules may
indicate their role in tissue interactions. For example, at the
time of odontoblast differentiation, there is a marked increase
in the amount of fibronectin in the dental basement membrane
(Thesleff *et al*, 1979), whereas no changes were observed in the
distribution of three other basement membrane components, i.e.
type IV collagen, laminin and a basement membrane proteoglycan
(Thesleff *et al*, 1980; Figure 11). Fibronectin promotes the
attachment of cells to collagen *in vitro* (Klebe, 1974), and it
probably interacts with other ECM macromolecules, e.g. GAGs
(Perkins *et al*, 1979). Thus it is possible that this

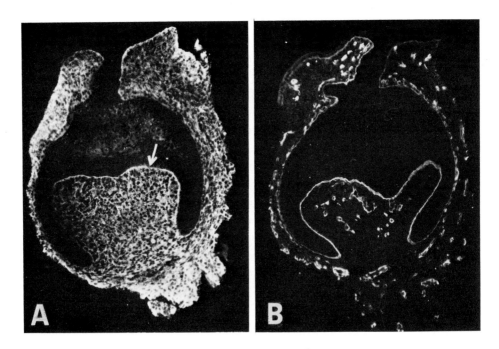

Figure 11: Localization by immunofluorescence of fibronectin (A) and type IV
collagen (B) in mouse molar tooth germs. The developmental stage is comparable
to the left part of Figure 9, i.e. the differentiation of odontoblasts is not
yet evident. Type IV collagen is evenly distributed in all basement membranes,
whereas fibronectin is found particularly abundantly under the enamel
epithelium (arrow).

glycoprotein plays a role in the alignment of mesenchymal cells under the epithelial basement membrane and their differentiation into odontoblasts.

Experimental analysis of the involvement of various basement membrane molecules in interactive events is, however, tricky, as already pointed out by Hay and Revel (1969). In most *in vitro* studies, the interacting tissue components are separated by an enzyme treatment, which also removes the basement membrane. As a consequence, these systems may in part measure the ability of a tissue to reconstruct a basement membrane. Furthermore, the results of studies in which the tissue interactions were inhibited by affecting one of the basement membrane molecules are difficult to interpret, because the interference with one of the components probably affects the overall assembly of the ECM. For example, the differentiation of odontoblasts can be inhibited by inter-ference with collagen deposition (Review: Kollar, 1978), with GAG sulfation (Hurmerinta *et al*, 1979, 1980) or with protein glycosylation (Thesleff & Pratt, 1980). Thus, even if one of these components was really inductive, establishment or proof by inhibitor experiments is difficult. It is also possible that no single macromolecule is inductive in this system, or that the three dimensional assembly of the basement membrane, created by interactions of the various macromolecules, is necessary.

The basement membranes of developing epithelial organs are particularly rich in GAG, compared to mature organs (Banerjee & Bernfield, 1976). A function for basal laminar GAG in morpho-genesis has been proposed by Bernfield (Review: Bernfield *et al*, 1973). In the basement membrane of the branching salivary gland epithelium, there is a rapid turnover of GAG at the distal aspects of the lobules, where the branching occurs. In contrast, the accumulation of GAG is found within the clefts. In the absence of mesenchyme, no branching occurs. The mesenchyme determines the differential turnover rates of GAG (Bernfield & Banerjee, 1978; Figure 12), by promoting the degradation of GAG. This effect of the mesenchyme may be mediated by the close contacts between epithelial and mesenchymal cells, since, at the lobular tips where GAG rapid turnover occurs, the epithelium grows quickly and the basal lamina is interrupted (Cutler & Chaudry, 1973).

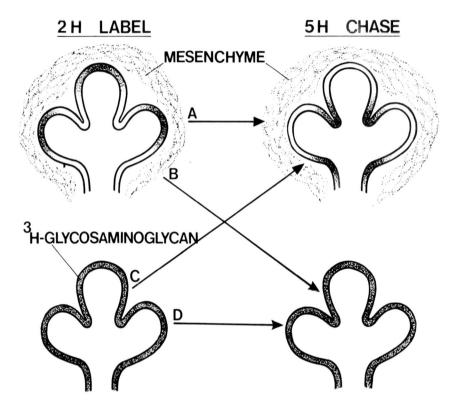

2 H   LABEL   5 H   CHASE

MESENCHYME

A

B

³H-GLYCOSAMINOGLYCAN

C

D

Figure 12: Effect of mesenchyme on the autoradiographic distribution of GAG label in the basement membrane of the embryonic submandibular gland. The labelling of the intact glands (upper left) and of collagenase-isolated epithelia (lower left) was for 2 h. The epithelia were then chased in the presence (A,C) or absence (B,D) of the mesenchyme for 5 h (right). The rapid turnover of GAG at the distal aspects of the lobules undergoing branching is dependent on the presence of the mesenchyme (After Bernfield & Banerjee, 1978).

Although it is believed generally that the ECM mediates morphogenetic tissue interactions in some systems, only stimulatory effects of the ECM molecules have so far been demonstrated. In no systems have the ECM molecules been shown to mediate an instructive type of interaction. At the developmental stage studied, the somitic mesenchyme undergoes chondrogenesis also when cultured alone, although the ECM causes enhancement in the rate of differentiation (Figure 6). Similarly, the corneal epithelium can synthesize collagen in the absence of the ECM, which, again, acts by stimulating this synthesis (Figure 8). The mesenchymal cells in the tooth germ have an "odontoblastic bias" (Kollar & Baird, 1970), but they do not differentiate, nor do they produce predentin in the absence of epithelium. However, even this proposed cell-matrix interaction is permissive in nature.

4. CELL MEDIATED INTERACTIONS

The hypothesis that inductive interactions require close cell contacts was suggested by Spemann (1901, 1938). This view was abandoned when it was found that many inductive events could take place through membrane filters (Grobstein, 1956b). With the methodology then available, no cytoplasmic material could be detected in the filter pores, and it was therefore postulated that extracellular components were the mediators in many of the secondary inductive events. Subsequent reports demonstrating cytoplasmic penetration into the filter pores and the ensuing suggestion that these close cell contacts might be related to induction were therefore somewhat unexpected (Saxen, 1972; Wartiovaara *et al*, 1974). Cell contact mediated induction has been described in the kidney model-system, and the occurrence of close contacts between the interacting epithelial and mesenchymal cells has also been reported in the developing tooth (Kallenbach, 1971; Slavkin & Bringas, 1976), in the duodenal mucosa (Mathan *et al*, 1972), in the lung (Bluemink *et al*, 1976), and in the liver (Rifkind, 1969), but the importance of the contacts in these cell interactions has not been established. Intercellular epithelio-mesenchymal contacts are also found in the developing sub-mandibular gland (Cutler & Chaudry, 1973), and Cutler (1977) has suggested recently that these contacts are an integral part of the developmental sequence. In what follows, the kidney will be taken as an example of a model-system where the inductive event seems to operate through cell contacts.

4.1 INDUCTION OF KIDNEY TUBULES

The mammalian metanephric kidney develops from two interacting tissue components, the epithelial ureter bud and the surrounding mesenchyme. Neither of the components differentiates if cultured alone, but if recombined, kidney tubules form in the mesenchyme and the ureter bud branches to form the collecting ducts

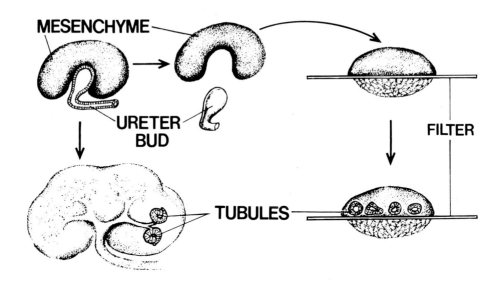

Figure 13: Scheme of the transfilter induction of the metanephric mesenchyme (Details: Grobstein, 1956; Wartiovaara *et al*, 1974).

(Grobstein, 1953, 1955b). The mesenchyme can be induced *in vitro* by use of a transfilter technique (Figure 13) that allows exact timing of the events (Grobstein, 1956b, 1957, 1967; Saxen & Saksela, 1971; Saxen & Lehtonen, 1978). The morphological response to induction resembles closely the *in vivo* events: epithelial tubules with a basement membrane develop, and subsequently different segments of the nephron form (Ekblom *et al*, 1980a,b).

Transmission mechanisms

Although long range diffusion might be the mode of transmission in primary induction (section 2), it does not seem likely to be a general mechanism operative in the kidney model-system. This has been shown in experiments comparing diffusion rates and the induction time. The minimum time required for the completion of tubule induction has been shown in transfilter experiments to be in the order of 12 to 24 hours (Nordling *et al*, 1971; Saxen & Lehtonen, 1978). Beyond this time, the responding mesenchyme differentiatiates even after the removal of the inductor. An additional filter placed between the tissues prolongs the minimum induction time to 30 hours. If the inductive signal were trans-ferred by diffusion, one would expect a similar delay in the diffusion rate of the signal. Compounds of different molecular sizes were found to diffuse rapidly through the filter, and the rate was not markedly changed by the interposition of an additional filter. These calculations indicate therefore that long-range diffusion is not likely to account for the time required for induction (Nordling *et al*, 1971). The possibility that cytoplasmic material could be present in the filter pores was therefore reinvestigated. Nuclepore filters with straight pores of various diameters were tested but only those with pore sizes larger than 0.1μm allowed the passage of induction and subsequent tubulogenesis. Modifications of the fixation methods for electron microscopy allowed clear demonstration that cell processes were present only in these filters of 0.1μm or greater (Wartiovaara *et al*, 1972, 1974), in contrast to previous findings with Millipore filters (Figure 14; Grobstein & Dalton, 1957). Moreover, explants cultivated for the minimum induction time also revealed cytoplasmic processes in the filter pores. Matrix material was not detected in the pores by electron microscopy. Thus, a correlation was demonstrated between close cell contacts and induction (Wartiovaara *et al*, 1974; Lehtonen, 1976). Further evidence favouring the hypothesis that intimate cell contacts mediate induction was obtained in similar types of experiment using different inductors. Differences between the transfilter inductive capacity of various inductors were shown, and could be correlated with their ability to send cytoplasmic processes into the filter pores (Saxen *et al*, 1976b).

The epithelio-mesenchymal interface

It could be argued that the *in vitro* results favouring the hypothesis of close cell contacts in induction is not relevant to the *in vivo* situation. Therefore, a thorough examination of the epithelio-mesenchymal interface between the natural inductor and the mesenchyme was warranted. This interface has been considered

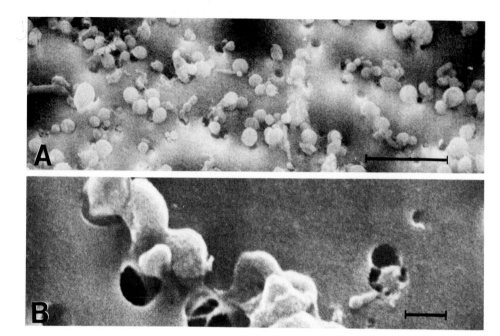

Figure 14: Scanning electron micrographs of the upper surface of a Nuclepore
filter after a 24 h cultivation of the inductor (spinal cord) cemented on the
lower surface. Cell processes have penetrated the filter (Courtesy
Dr. E. Lehtonen) (Scale A, 10μm; B, 1.0μm).

to consist of a basement membrane in most organs. Lehtonen (1975),
however, showed that the space between the epithelium and the
mesenchyme in the kidney was at places less than 10nm, and that
discontinuities of the basement membrane occurred at the
inductively active tips. Electron microscopy showed that the
cells were only partly separated by a ruthenium red positive coat,
and close appositions of the epithelial and mesenchymal cells were
found in places (Figure 15A; Lehtonen, 1975, 1976).

These findings have been confirmed by using other methods.
The ruthenium red staining is a nonspecific cationic probe, staining
a variety of substances such as glycosaminoglycans (Martinez-
Palomo, 1970). Lectins can be used for staining saccharide
moieties and their specificity is better known (Nicolson, 1974).
In the kidney, many lectins give essentially the same pattern as
the ruthenium red staining (Figure 15B): at inductively active
sites, the interspace seems to have less saccharide moieties.

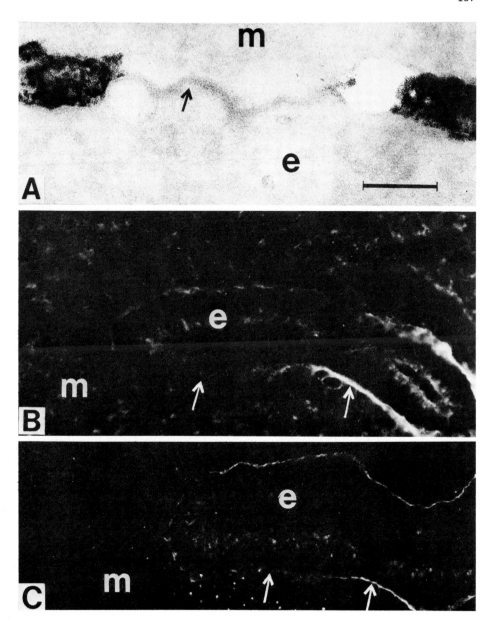

Figure 15: Epithelio-mesenchymal (e-m) interface in intact kidney rudiments of 12-day-old mouse embryos. (A) ruthenium red staining between the interacting cells, intimate cell contacts arrowed (Lehtonen, 1976). (Scale 0.2μm). (B) fluorescent wheat germ agglutinin (WGA) binds to the stalk (right arrow) but not around the tip (left arrow). (C) laminin immunofluorescence reveals discontinuity of the basement membrane at the tip (left). Note also laminin spots in the induced mesenchyme (Ekblom et al, 1980a).

Staining with antibodies against basement membrane proteins such
as laminin (Timpl *et al*, 1979b) also reveals discontinuities at
the inductively active tips (Figure 15C) (Ekblom *et al*, 1980a).
These *in vivo* findings are therefore compatible with the hypothesis
that induction operates through cell contacts in the kidney.

4.2   *MECHANISMS OF CONTACT-MEDIATED INDUCTION*

Proposals that close cell contacts are required in induction
naturally raise further questions on the nature of this "contact".
An understanding of the molecular mechanism in the transmission
of the signals requires a more thorough characterization of the
surface structures in the interface. One should also have a
reasonably adequate view of the nature of the response. Should we
look for intracellular or extracellular compounds, integral
membrane glycoproteins, surface associated glycosaminoglycans,
ultrastructurally detectable gap junctions, or should we focus on
receptors such as adenylate cyclase type molecules?   Three different types of contact mediated interactions can be postulated
(Figure 16): short-range diffusion of the signal substances
(Weiss & Nir, 1979), transfer through specialized intercellular
junctions (gap junctions) known to allow the passage of small
molecules (Gilula *et al*, 1972; Lo, this volume), or cell surface
components (glycoproteins, glycolipids, GAGs). In many different
types of cell-cell interactions in embryogenesis, interaction
between surface glycoconjugates has been suggested to be of
importance (Reviews: Roseman, 1974; Moscona, 1976), and there are
several well documented situations where glycoproteins or other
surface receptors are required. This is particularly evident in
many interactions that require cell adhesion and recognition
between like cells. Whether this type of interaction could also
be involved in induction between heterotypic cell populations is
not yet known.

A general model that might be applicable to many interactions
has been proposed by Edelman (1976). He argues that lateral
mobility in the lipid bilayer of the surface receptors provides
a basis for hypotheses on molecular mechanisms mediating cell
recognition and growth control. He further suggests that cell-
cell interactions might be coordinated by an assembly of inter-

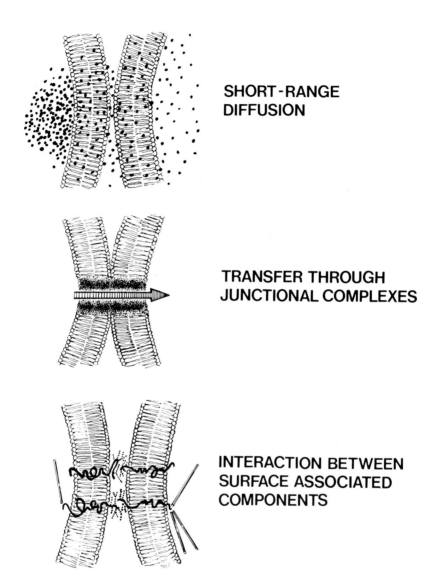

SHORT-RANGE DIFFUSION

TRANSFER THROUGH JUNCTIONAL COMPLEXES

INTERACTION BETWEEN SURFACE ASSOCIATED COMPONENTS

Figure 16: Alternative modes of contact mediated induction.

acting macromolecules consisting of cell surface receptors, transmembrane components, and submembranous fibrillar structures. Another type of surface interaction based on enzyme-substrate interaction has been advanced by Roseman, Roth and co-workers (Reviews: Roseman, 1974; Roth *et al*, 1977). The activity of cell surface glycosyltransferases shows a clear correlation with cell migration in some situations (Roth, 1973; Shur, 1977), and in many adhesive recognition phenomena similar types of enzymes have been implicated (Roth *et al*, 1971; McClean & Bosmann, 1975; Durr *et al*, 1977). This type of enzyme-substrate interaction could well be operative in inductive events, but the specific enzyme involved could differ in each interaction. The activity of hyaluronidase seems to correlate with morphogenesis in the limb (Toole & Gross, 1971). Other similar types of enzyme-triggered events on the surface could be involved in the kidney model-system.

The cell surface can be altered with a variety of techniques. These types of study have again suggested the cell surface to be a mediator of the kidney tubule induction. They do not, however, necessarily distinguish between the alternatives presented above. Charged polymers such as heparin are known to alter surface properties (Nordling *et al*, 1965). Therefore we investigated the effect of various polyanions on the induction of kidney tubules. These molecules were found to inhibit induction, and the effect was a function of the charge density of the polyanions used (Ekblom *et al*, 1978). It is of interest to note that these polyanions have been shown to modulate the activity of adenylate cyclase (Salomon *et al*, 1978; Amsterdam *et al*, 1978).

Experiments with inhibitors of protein glycosylation have suggested a role for glycoproteins during the induction period. Two inhibitors of protein glycosylation were used: 6-diazo-5-oxo-norleucine (Ekblom *et al*, 1979) and tunicamycin (Ekblom *et al*, 1979). Both agents were found to inhibit induction, but they did not prevent the subsequent stages of morphogenesis, nor did they prevent passage of cell processes through the filter pores. No general toxicity was observed. The effect of the drugs was reversible. The decrease in protein glycosylation was found to correlate with the inhibition of induction (Figure 17). It is

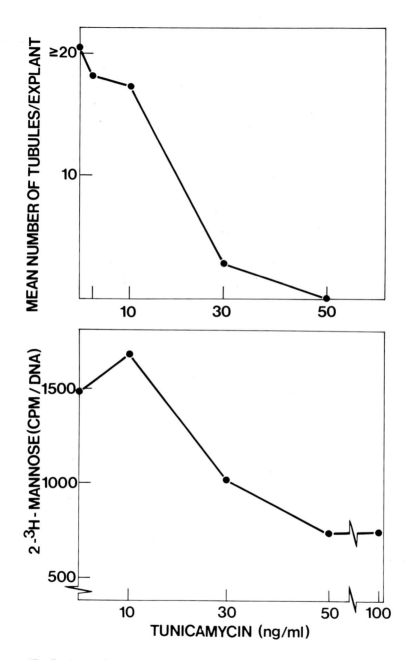

Figure 17: Tunicamycin was present during a 24 h induction period; tunicamycin and the inductor were then removed and the mesenchyme was cultured for 48 h. (A) The number of tubules formed were counted. (B) 2-^{3}H-mannose (given at 22-24 h) incorporation into spinal cord proteins was assessed (Ekblom *et al*, 1979a).

therefore logical to assume that newly synthesized glycoproteins
are required during the induction period. As both the inductor
and the mesenchyme were exposed to the inhibitors, impaired
glycosylation could have blocked induction by affecting either
the cells in the inductor tissue or the kidney mesenchyme. The
molecules involved in induction are still unknown, but the
results suggest glycoconjugates, possibly glycoproteins, on the
cell surface.

4.3   INDUCTION AND MORPHOGENESIS

The results in the kidney model-system speak in favour of
induction mechanisms based on intimate associations between the
interacting cells, making matrix interactions an unlikely
mechanism in the transmission of the inductive signal. This does
not, however, exclude the possibility that other events in kidney
development are matrix mediated. The different phases in kidney
development can be divided into the induction period when no
morphological signs of differentiation can be seen, the aggregation period when the induced cells form aggregates, and the
tubulogenic period when tubules with a lumen and a basement
membrane form. All the morphological events occur autonomously
without the inductor once the induction has been completed (Saxen
& Lehtonen, 1978). It means that the induced cells have acquired
features that distinguish them from uninduced cells. A
molecularly defined difference between these cells might provide
a link between morphogenesis and induction. Recently, one such
difference has been found. During the induction period, a glycoprotein specific for basement membranes appears in the mesenchyme,
apparently as a consequence of induction (Figure 15C). This
protein, laminin (Timpl et al, 1979b), is seen prominently in the
cell aggregates. During tubulogenesis it becomes confined to the
developing basement membranes (Figure 18; Ekblom et al, 1980a).
Antigens specific for the luminal side of the tubules (Miettinen
& Linder, 1976) appear later, after the formation of the basement
membrane (Figure 19; Ekblom et al, 1980b). The results show that
the ECM changes as a response to induction. This occurs prior to
morphogenesis, and the distribution of laminin in the later stages
makes it reasonable to assume that this change is significant in
morphogenesis. Other basement membrane components might appear in

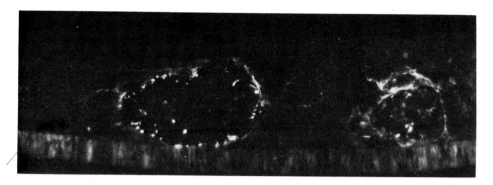

Figure 18: Immunofluorescent localization of laminin in developing tubules of mesenchyme cultured for 48 h. Note staining in the developing basement membrane and also intracellularly. The inductor was removed at 24 h. (Ekblom *et al*, 1980a).

a similiar way. It also seems that mesenchymal type matrix proteins (interstitial collagens, fibronectin) disappear during the induction period (Ekblom *et al*, 1980c).

The present working hypothesis in the kidney model-system can therefore be summarized as follows. The inductive interaction triggering the metanephrogenic mesenchyme requires intimate cell contacts. These contacts form rapidly. Protein glycosylation is required during the induction period. During this period, changes occur in the ECM: the mesenchymal matrix components disappear, and basement membrane components appear. Once these matrix changes have been completed, the inductor tissue is no longer

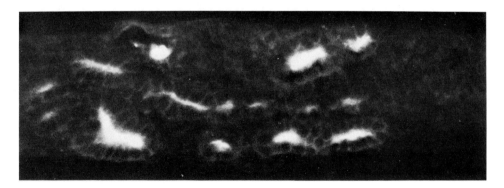

Figure 19: Immunofluorescent localization of brush border antigens in tubules of tissues cultured for 96 h. The inductor was removed at 24 h. (Ekblom *et al*, 1980b).

needed. The basement membrane forms, and after the development of the linear basement membrane the cells build up the luminal side of the tubules.

5. CONCLUDING REMARKS

Our present, still fragmentary knowledge of the localization and transmission of morphogenetic signal substances shows that different interactive events have different mechanisms. No common denominators have been found so far, and the diverging mechanisms suggest that the broad term "morphogenetic cell interactions" covers a wide variety of developmental control events. The signals in these interactions differ in their biological nature; some are "directive" and others "permissive" (Saxen, 1977). Consequently, future research on these developmentally significant processes should not necessarily focus on a search for unifying concepts. Premature generalizations may obscure the details and unique features of individual events.

REFERENCES

AMSTERDAM, A., RECHES, A., AMIR, Y., MINZ, Y. & SALOMON, Y. (1978) Modulation of adenylate cyclase activity by sulfated glycosaminoglycans. II. Effects of mucopolysaccharides and dextran sulfate of the activity of adenylate cyclase derived from various tissues. Biochim. Biophys. Acta 544, 273-283.

BANERJEE, S.D. & BERNFIELD, M.R. (1976) Pattern of deposition and loss of basal laminar glycosaminoglycan during epithelial morphogenesis. J. Cell Biol. 70, 111a.

BARTH, L.G. & BARTH, L.J. (1968) The role of sodium chloride in the process of induction by lithium chloride in cells of the Rana pipiens gastrula. J. Embryol. exp. Morph. 19, 387-396.

BARTH, L.G. & BARTH, L.J. (1972) ^{22}Sodium and ^{45}calcium uptake during embryonic induction in Rana pipiens. Dev. Biol. 28, 18-34.

BAUTZMANN, H., HOLTFRETER, J., SPEMANN, H. & MANGOLD, O. (1932) Versuche zur Analyse der Induktionsmittel in der Embryonalentwicklung. Naturwissenschaften 20, 971-974.

BERNFIELD, M.R. & BANERJEE, S.D. (1978) The basal lamina in epithelial mesenchymal morphogenetic interactions. In: N.A. Kefalides (Ed.), Biology and Chemistry of Basement Membranes, New York-London, pp. 137-148.

BERNFIELD, M.R., COHN, R.H. & BANERJEE, S.D. (1973) Glycosaminoglycans and epithelial organ formation. Amer. Zool. 13, 1067-1083.

BLUEMINK, J.G., van MAURIK, P. & LAWSON, K.A. (1976) Intimate cell contacts at the epithelial/mesenchymal interphase in embryonic mouse lung. J. Ultrastruct. Res. 55, 257-270.

CLAYTON, R.M. (1977) Convergence and divergence in lens cell differentiation: Regulation of the formation and specific content of lens fibre cells. In: Stem Cells and Tissue Homeostasis, Cambridge University Press, London.

COHEN, A. (1961) Electron microscopic observations of the developing mouse eye. I. Basement membranes during early development and lens formation. Dev. Biol. 3, 297-316.

COULOMBRE, A.J. & COULOMBRE, J.L. (1971) The development of the structural and optical properties of the cornea. In: G.K. Smelser (Ed.), The Structure of the Eye, Academic Press, New York, pp. 405-420.

CUTLER, L.S. (1977) Intercellular contacts at the epithelial-mesenchymal interface of the developing rat submandibular in vitro. J. Embryol. exp. Morph. 39, 71-77.

CUTLER, L.S. & CHAUDHRY, A.P. (1973) Intercellular contacts at the epithelial-mesenchymal interface during the prenatal development of the rat submandibular gland. Dev. Biol. 33, 229-240.

DODSON, J.W. & HAY, E.D. (1971) Secretion of collagenous stroma by isolated epithelium grown in vitro. Exp. Cell Res. 65, 215-220.

DURR, R., SHUR, B. & ROTH, S. (1977) Surface sialyltransferases on mouse eggs and sperm. Nature (Lond.) 265, 547-548.

EAKIN, R.M. & LEHMAN, F.E. (1957) An electromicroscopic study of developing amphibian ectoderm. Arch. EntwMech. Org. 150, 177-198.

EDELMAN, G.M. (1976) Surface modulation in cell recognition and cell growth. Science 192, 218-226.

EKBLOM, P., NORDLING, S. & SAXEN, L. (1978) Inhibition of kidney tubule induction by charged polymers. Cell Differ. 7, 345-353.

EKBLOM, P., NORDLING, S., SAXEN, L., RASILO, M.-L. & RENKONEN, O. (1979a) Cell interactions leading to kidney tubule determination are tunicamycin sensitive. Cell Differ. 8, 347-352.

EKBLOM, P., LASH, J.W., LEHTONEN, E., NORDLING, S. & SAXEN, L. (1979b) Inhibition of morphogenetic cell interactions by 6-diazo-5-oxo-norleucine (DON). Exp. Cell Res. 121, 121-126.

EKBLOM, P., ALITALO, K., VAHERI, A., TIMPL, R. & SAXEN, L. (1980a) Induction of a basement membrane glycoprotein in embryonic kidney; possible role of laminin in morphogenesis. Proc. Natl. Acad. Sci. USA 77, 485-489.

EKBLOM, P., MIETTINEN, A. & SAXEN, L. (1980b) Induction of brush border antigens of the proximal tubule in the developing kidney. Dev. Biol. 74, 263-274.

EKBLOM, P., SAXEN, L. & TIMPL, R. (1980c) Shift in collagen type as an early response to induction of the metanephric mesenchyme, submitted.

FOIDART, J.-M., BERE, E.W., YAAR, M., RENNARD, S., GULLINO, M. & MARTIN, G.R. (1980) Distribution and immunoelectron microscopic localization of laminin, a non-collagenous basement membrane glycoprotein. Lab. Invest., in press.

GAUNT, W.A. & MILES, A.E.W. (1967) Fundamental aspects of tooth morphogenesis. In: A.E.W. Miles (Ed.), Structural and Chemical Organization of Teeth, Vol. 1, Academic Press, New York, pp. 151-197.

GILULA, G.B., REEVES, O.R. & STEINBACH, A. (1972) Metabolic coupling, ionic coupling and cell contacts. Nature (London) 235, 262-265.

GROBSTEIN, C. (1953) Morphogenetic interaction between embryonic mouse tissues separated by a membrane filter. Nature (Lond.) 172, 869-871.

GROBSTEIN, C. (1955a) Tissue interaction in the morphogenesis of mouse embryonic rudiments in vitro. In: D. Rudnick (Ed.), Aspects of Synthesis and Order in Growth, Princeton University Press, Princeton, pp. 233-256.

GROBSTEIN, C. (1955b) Inductive interaction in the development of the mouse metanephros. J. Exp. Zool. 130, 319-340.

GROBSTEIN, C. (1956a) Inductive tissue interactions in development. Adv. Cancer Res. 4, 187-236.

GROBSTEIN, C. (1956b) Transfilter induction of tubules in mouse metanephric mesenchyme. Exp. Cell Res. 10, 424-440.

GROBSTEIN, C. (1957) Some transmission characteristics of the tubule-inducing influence on mouse metanephrogenic mesenchyme. Exp. Cell Res. 13, 575-587.

GROBSTEIN, C. (1967) Mechanism of organogenetic tissue interaction. Natl. Cancer Inst. Monogr. 26, 279-299.

GROBSTEIN, C. & DALTON, A.J. (1957) Kidney tubule induction in mouse metanephrogenic mesenchyme without cytoplasmic contact. J. Exp. Zool. 135, 57-73.

GRUNZ, H. & STAUBACH, J. (1979) Cell contacts between chorda-mesoderm and the overlaying neuroectoderm (presumptive central nervous system) during the period of primary embryonic induction in amphibians. Differentiation 14, 59-65.

GRUNZ, H. & TIEDEMANN, H. (1977) Influence of cyclic nucleotides on amphibian ectoderm. Wilhelm Roux' Archives 181, 261-265.

HASSELL, J.R., ROBEY, P.G., BARRACH, H.J., WILCZEK, J., RENNARD, S.I. & MARTIN, G.R. (1979) Basement membrane proteoglycans. In: 5th International Symposium on Glycoconjugates, Kiel, FDR. Eds. R. Schauer et al., pp.728-729.

HAY, E. (1978) Embryonic induction and tissue interaction during morphogenesis. In: J.W. Littlefield and J. de Grouchy (Eds.), Birth Defects, Exerpta Medica, Amsterdam-Oxford, pp. 126-140.

HAY, E.D. (1977) Interaction between the cell surface and extracellular matrix in corneal development. In: J.W. Lash and M.M. Burger (Eds.), Raven Press, New York, pp. 115-137.

HAY, E.D. & DODSON, J.W. (1973) Secretion of collagen by corneal epithelium. I. Morphology of the collagenous produced by isolated epithelia grown on frozen-killed lens. J. Cell Biol. 57, 190-213.

HAY, E.D. & MEIER, S. (1974) Glycosaminoglycan synthesis by embryonic inductors: Neural tube, notochord and lens. J. Cell Biol. 62, 889-898.

HAY, E.D. & MEIER, S. (1976) Stimulation of corneal differentiation by interaction between cell surface and extracellular matrix. II. Further studies on the nature and site of transfilter "induction". Dev. Biol. 52, 141-157.

HAY, E.D. & REVEL, J.P. (1969) Fine structure of the developing avian cornea. In: A. Wolsky and P.S. Chen (Eds.), Monographs in Developmental Biology, Vol. 1, Karger, Basel.

HENDRIX, R.W. & ZWAAN, J. (1974) Changes in the glycoprotein concentration of the extracellular matrix between lens and optic vesicle associated with early lens differentiation. Differentiation 2, 357-362.

HOLTFRETER, J. (1955) Studies on the diffucibility, toxicity and pathogenic properties of "inductive" agents derived from dead tissues. Exp. Cell Res. Suppl. 3, 188-209.

HOLTZER, H. & DETWILER, A.R. (1953) An experimental analysis of the development of the spinal column. III. Induction of skeletogenous cells. J. Exp. Zool. 123, 335-366.

HUNT, H.H. (1961) A study of the fine structure of the optic vesicle and lens placode of the chick embryo during induction. Dev. Biol. 3, 175-209.

HURMERINTA, K., THESLEFF, I. & SAXEN, L. (1979) Inhibition of tooth germ differentiation in vitro by diazo-oxo-norleucine (DON). J. Embryol. exp. Morph. 50, 99-109.

HURMERINTA, K., THESLEFF, I. & SAXEN, L. (1980) In vitro inhibition of odontoblast differentiation by vitamin A. Arch. oral Biol., in press.

JACOBSON, A.G. (1958) The roles of neural and non-neural tissues in lens induction. J. Exp. Zool. 139, 525-558.

JOHNEN, A.G. (1970) Der Einfluss von Li-und SCN-Ionen auf die Differenzierungsleistung des Ambystoma-Ektoderms und ihre Veranderung bei kombinierter Einwirkung beider Ionen. Wilhelm Roux' Archiv 165, 150-162.

KALLENBACH, E. (1971) Electron microscopy of the differentiating rat incisor ameloblast. J. Ultrastruct. Res. 35, 508-531.

KANWAR, Y.S. & FARQUHAR, M.G. (1979) Presence of heparan sulfate in the glomerular basement membrane. Proc. Natl. Acad. Sci. USA 76, 1303-1307.

KARKINEN-JAASKELAINEN, M. (1978a) Permissive and directive interactions in lens induction. J. Embryol. exp. Morph. 44, 167-179.

KARKINEN-JAASKELAINEN, M. (1978b) Transfilter lens induction in avian embryo. Differentiation 12, 31-37.

KASKA, D. & TRIPLETT, E.L. (1980) Inductive interaction of amphibian chordamesoderm and ectoderm. Induction of ectodermal tyrosine oxidase synthesis by chordamesoderm conditioned media., submitted.

KEFALIDES, N.A. (Ed.) (1978) Biology and Chemistry of Basement Membranes, Academic Press, New York.

KELLER, R.E. & SCHOENWOLF, G.C. (1977) An SEM study of cellular morphology, contact, and arrangement, as related to gastrulation in Xenopus laevis. Wilhelm Roux' Archives 182, 165-186.

KLEBE, R.J. (1974) Isolation of a collagen-dependent cell attachment factor. Nature (Lond.) 250, 248-251.

KOLLAR, E.J. (1978) The role of collagen during tooth morphogenesis: Some genetic implications. In: P.M. Butler and K.A. Joysey (Eds.), Development, Function and Evolution of Teeth, Academic Press, London, pp. 1-12.

KOLLAR, E.J. & BAIRD, G. (1970) Tissue interactions in embryonic mouse tooth germs. II. The inductive role of the dental papilla. J. Embryol. exp. Morph. 24, 173-186.

KONIGSBERG, I.R. & HAUSCHKA, S.D. (1965) Cell and tissue interactions in the reproduction of cell type. In: M. Locke (Ed.), Reproduction: Molecular, Subcellular, and Cellular, Academic Press, New York, pp. 243-290.

KOSHER, R.A. (1976) Inhibition of "spontaneous", notochord-induced, and collagen-induced in vitro somite chondrogenesis by cyclic AMP derivatives and theophylline. Dev. Biol. 53, 265-276.

KOSHER, R.A. & LASH, J.W. (1975) Notochordal stimulation of in vitro somite chondrogenesis before and after enzymatic removal of perinotochordal materials. Dev. Biol. 42, 362-378.

KRATOCHWILL, K. (1972) Tissue interaction during embryonic development. In: D. Tarin (Ed.), Tissue Interactions and Carcinogenesis, Academic Press, London, pp. 1-47.

LANDSTROM, U. & LØVTRUP, S. (1977) Is heparan sulphate the agent of the amphibian inductor? Acta Embryol. Exp. 1977, 171-178.

LASH, J.W. & VASAN, N.S. (1977) Tissue interactions and extracellular matrix components. In: J.W. Lash and M.M. Burger (Eds.), Cell and Tissue Interactions, Raven Press, pp. 101-113.

LASH, J.W. & VASAN, N.S. (1978) Somite chondrogenesis in vitro. Stimulation by extracellular matrix components. Dev. Biol. 66, 151-171.

LEHTONEN, E. (1975) Epithelio-mesenchymal interface during mouse kidney tubule induction in vivo. J. Embryol. exp. Morph. 34, 695-705.

LEHTONEN, E. (1976) Transmission of signals in embryonic induction. Med. Biol. 54, 108-158.

LESOT, H., VON DER MARK, K. & RUCH, J.V. (1978) Localization par immuno-fluorescence des types de collagene syntheses par l'ebauche dentaire chez l'embryon de Souris. S. R. Acad. Sci. Paris 286, 765-768.

LINDER, E., VAHERI, A., RUOSLAHTI, E. & WARTIOVAARA, J. (1975) Distribution of fibroblast surface antigen in the developing chick embryo. J. exp. Med. 142, 41-49.

LINSENMAYER, T.F., TRELSTAD, R.L. & GROSS, J. (1973) The collagen of chick embryonic notochord. Biochem. Biophys. Res. Commun. 53, 39-45.

LOPASHOV, G.V. & STROEVA, O.G. (1961) Morphogenesis of the vertebrate eye. Advanc. Morphogenes. 1, 331-377.

MARTINEZ-PALOMO, A. (1970) The surface coats of animal cells. Int. Rev. Cytol. 29, 29-75.

MASUI, Y. (1959) Induction of neural structures under the influence of lithium chloride. Annot. zool. jap. 32, 23-30.

MATHAN, M., HERMOS, J.A. & TRIER, J.S. (1972) Structural features of the epithelio-mesenchymal interface of rat duodenal mucosa during development. J. Cell Biol. 52, 577-588.

McCLEAN, R.J. & BOSMANN, H.B. (1975) Enchanchement of glycosylatransferase ectoenzyme systems during Chlamydomonas gametic contact. Proc. Natl. Acad. Sci. USA 72, 310-313.

McMAHON, D. (1974) Chemical messengers in development: A hypothesis. Science 185, 1012-1021.

McMAHON, D. & WEST, C. (1976) Transduction of positional information during development. In: G. Poste and G.L. Nicolson (Eds.), The Cell Surface in Animal Embryogenesis and Development, North-Holland Publ. Co., Amsterdam, pp. 449-493.

MEIER, S. & HAY, E.D. (1974a) Control of corneal differentiation by extracellular materials. Collagen as a promoter and stabilizer of epithelial stroma production. Dev. Biol. 38, 249-270.

MEIER, S. & HAY, E.D. (1974b) Stimulation of extracellular matrix synthesis in the developing cornea by glycosaminoglycans. Proc. Natl. Acad. Sci. USA 71, 2310-2313.

MEYER, J.M., FABRE, M., STAUBLI, A. & RUCH, J.V. (1977) Relations cellulaires au cours de l'odontogenese. J. Biol. buccale 5, 107-109.

MIETTINEN, & LINDER, E. (1976) Membrane antigens shared by renal proximal tubules and other epithelia associated with absorption and excretion. Clin. Exp. Immunol. 23, 568-577.

MINUTH, W.W. (1978) Mesodermalizing of Amphibian gastrula ectoderm in transfilter experiments. Med. Biol. 56, 349-354.

MIZUNO, T. (1972) Lens differentiation in vitro in the absence of optic vesicle in the epiblast of chick blastoderm under the influence of skin dermis. J. Embryol. exp. Morph. 28, 117-132.

MOSCONA, A. (1976) Cell recognition in embryonic morphogenesis and the problem of neuronal specificities. In: S. Barondes (Ed.), "Neuronal Recognition", Plenum Press, New York, pp. 205-226.

MUTHUKKARUPPAN, V. (1965) Inductive tissue interaction in the development of the mouse lens in vitro. J. Exp. Zool. 159, 269-288.

NICOLSON, G.L. (1974) The interactions of lectins with animal cell surfaces. Int. Rev. Cytol. 89-190.

NIEUWKOOP, P.D. (1973) The "organization center" of the amphibian embryo: its origin, spatial organization, and morphogenetic action. Advanc. Morphogenes. 10, 1-39.

NORDLING, S., VAHERI, A., SAXEN, E. & PENTTINEN, K. (1965) The effects of anionic polymers on cell attachment and growth behaviour, with a note on a similarity in the effect of fresh human serum. Exp. Cell Res. 37, 406-419.

NORDLING, S., MIETTINEN, H., WARTIOVAARA, J. & SAXEN, L. (1971) Transmission and spread of embryonic induction. I. Temporal relationship in transfilter induction of kidney tubules in vitro. J. Embryol. exp. Morphol. 26, 231-252.

NYHOLM, M., SAXEN, L., TOIVONEN, S. & VAINIO, T. (1962) Electron microscopy of transfilter neural induction. Exp. Cell Res. 28, 209-212.

PERKINS, M.E., JI, T.H. & HYNES, R.O. (1979) Cross-linking of fibronectin to sulfated proteoglycans at the cell surface. Cell 16, 941-952.

RIFKIND, R.A., CHUI, D. & EPLER, H. (1969) An ultrastructural study of early morphogenetic events during the establishment of fetal hepatic erytropoiesis. J. Cell Biol. 40, 343-365.

ROSEMAN, S. (1974) Complex carbohydrates and intercellular adhesion. In: A. Moscona (Ed.) The Cell Surface in Development, John Wiley & Sons, New York, pp. 255-272.

ROTH, S. (1973) A molecular model for cell interactions. Quart. Rev. Biol. 48, 541-563.

ROTH, S., McGUIRE, E.J. & ROSEMAN, S. (1971) Evidence for cell surface glycosyltransferases. J. Cell Biol. 51, 536-547.

ROTH, S., SHUR, B.D. & DURR, R. (1977) A possible enzymatic basis for some cell recognition and migration phenomena in early embryogenesis. J.W. Lash and M.M. Burger (Eds.) Cell and Tissue Interactions, Raven Press, New York, pp. 209-232.

RUGGERI, A. (1972) Ultrastructural, histochemical and autoradiographic studies on the developing chick notochord. Z. Anat. Entwicklungsgesch. 138, 20-33.

SALOMON, Y., AMIR, Y., AZULAI, R. & AMSTERDAM, A. (1978) Modulation of adenylate cyclase activity by sulfated glycosaminoglycans. I. Inhibition by heparin of gonadotropin-stimulated ovarian adenylate cyclase. Biochim. Biophys. Acta 544, 262-272.

SAXEN, L. (1961) Transfilter neural induction of amphibian ectoderm. Dev. Biol. 3, 140-152.

SAXEN, L. (1963) The transmission of information during primary embryonic induction. In: R.J.C. Harris (Ed.), Biological Organization at Cellular and Supercellular Level, Academic Press, London, pp. 211-227.

SAXEN, L. (1972) Interactive mechanisms in morphogenesis. In: D. Tarin (Ed.), Tissue Interactions in Carcinogenesis, Academic Press, London, pp. 49-80.

SAXEN, L. (1977) Morphogenetic tissue interactions: An introduction. In: M. Karkinen-Jaaskelainen, L. Saxen and L. Weiss (Eds.), Cell Interactions in Differentiation, Academic Press, London, pp. 145-151.

SAXEN, L. & LEHTONEN, E. (1978) Transfilter induction of kidney tubules as a function of the extent and duration of intercellular contacts. J. Embryol. exp. Morph. 47, 97-109.

SAXEN, L. & SAKSELA, E. (1971) Transmission and spread of embryonic induction. II. Exclusion of an assimilatory transmission mechanism in kidney tubule induction. Exp. Cell Res. 66, 369-377.

SAXEN, L. & TOIVONEN, S. (1962) Primary Embryonic Induction, Logos Press, Academic Press, London and Prentice-Hall Inc., Englewood Cliffs, N.J.

SAXEN, L., TOIVONEN, S. & VAINIO, T. (1964) Initial stimulus and subsequent interactions in embryonic induction. J. Embryol. exp. Morph. 12, 333-338.

SAXEN, L., KARKINEN-JAASKELAINEN, M., LEHTONEN, E., NORDLING, S. & WARTIOVAARA, J. (1976a) Inductive tissue interactions. In: G. Poste and G.L. Nicolson (Eds.), The Cell Surface in Animal Embryogenesis and Development, North-Holland Publ. Co., Amsterdam, pp. 331-407.

SAXEN, L., LEHTONEN, E., KARKINEN-JAASKELAINEN, M., NORDLING, S. & WARTIOVAARA, J. (1976b) Are morphogenetic tissue interactions mediated by transmissible signal substances or through cell contacts? Nature 259, 662-663.

SHUR, B.D. (1977) Cell-surface glycosyl-transferases in gastrulating chick embryos. I. Temporally and spatially specific patterns of four endogenous glycosyltransferase activities. Dev. Biol. 58, 23-39.

SLAVKIN, H. & BRINGAS, P. (1976) Epithelial-mesenchymal interactions during odontogenesis. IV. Morphological evidence for direct heterotypic cell-cell contacts. Dev. Biol. 50, 428-442.

SLAVKIN, H.C. & GREULICH, R.C. (Eds.) (1975) Extracellular Matrix Influences on Gene Expression, Academic Press, New York.

SPEMANN, H. (1901) Uber Korrelationen in der Entwicklung des Auges. Anat. Anzeiger 19, Erganzungsheft, 61-79.

SPEMANN, H. (1938) Embryonic Development and Induction. Yale University Press, New Haven.

STRUDEL, G. (1953) L'influence morphogene du tube nerveux sur differen-ciation de la colone vertebrale. R.C. Soc. Biol. 147, 132-133.

THESLEFF, I. (1978) Role of the basement membrane in odontoblast differen-tiation. J. Biol. Buccale 6, 241-249.

THESLEFF, I., LEHTONEN, E., WARTIOVAARA, J. & SAXEN, L. (1977) Interference of tooth differentiation with interposed filters. Dev. Biol. 58, 197-203.

THESLEFF, I., LEHTONEN, E. & SAXEN, L. (1978) Basement membrane formation in transfilter tooth culture and its relation to odontoblast differentiation. Differentiation 10, 71-79.

THESLEFF, I., STENMAN, S., VAHERI, A. & TIMPL, R. (1979) Changes in matrix proteins, fibronectin and collagen, during differentiation of mouse tooth germ. Dev. Biol. 70, 116-126.

THESLEFF, I. & PRATT, R.M. (1980a) Tunicamycin inhibits mouse tooth morphogenesis and odontoblast differentiation in vitro., submitted.

THESLEFF, I., BARRACH, H.J., FOIDART, J.M., VAHERI, A., PRATT, R.M. & MARTIN, G.R. (1980b) Changes in the distribution of type IV collagen, laminin, proteoglycan and fibronectin during mouse tooth development., submitted.

TIEDEMANN, H. (1976) Pattern formation in early developmental stages of amphibian embryos. J. Embryol. exp. Morph. 35, 437-444.

TIEDEMANN, H. (1978) Chemical approach to the inducing agents. In: O. Nakamura and S. Toivonen (Eds.), Organizer - A milestone of a half-century from Spemann, Elsevier/North-Holland Biomed. Press, Amsterdam, pp. 91-117.

TIEDEMANN, H. & BORN, J. (1978) Biological activity of vegetalizing and neuralizing inducing factors after binding to BAC-cellulose and CNBr-Sepharose. Wilhelm Roux Archives 184, 285-299.

TIEDEMANN, H., TIEDEMANN, H., BORN, J. & KOCHER-BECKER, U. (1969) Wirkung von Sulfhydrylverbindungen auf embryonale Induktionsfaktoren. Wilhelm Roux Archives 163, 316-324.

TIMPL, R., GLANVILLE, R.W., WICK, G. & MARTIN, R. (1979a) Immunochemical study on basement membrane (type IV) collagens. Immunology 38, 109-116.

TIMPL, R., ROHDE, H., ROBEY, P.G., RENNARD, S.I., FOIDART, J.M. & MARTIN, G.R. (1979b) Laminin - a glycoprotein from basement membranes. J. Biol. Chem. 254, 9933-9937.

TOIVONEN, S. (1979) Transmission problem in primary induction. Differentiation, 15, 177-181.

TOIVONEN, S. & WARTIOVAARA, J. (1976) Mechanism of cell interaction during primary induction studied in transfilter experiments. Differentiation 5, 61-66.

TOIVONEN, S., TARIN, D. & SAXEN, L. (1976) The transmission of morphogenetic signals from amphibian mesoderm to ectoderm in primary induction. Differentiation 5, 49-55.

TOOLE, B.P. & GROSS, J. (1971) The extracellular matrix of regenerating newt limb: synthesis and removal of hyaluronate prior to differentiation. Dev. Biol. 25, 57-77.

WAHN, H.L., LIGHTBODY, L.E. & TCHEN, T.T. (1976) Induction of neural dif-ferentiation in cultures of amphibian undetermined presumptive epidermis by cyclic AMP derivates. Science 188, 366-369.

WARTIOVAARA, J., LEHTONEN, E., NORDLING, S. & SAXEN, L. (1972) Do membrane filters prevent cell contacts? Nature (Lond.) 238, 407-408.

WARTIOVAARA, J., NORDLING, S., LEHTONEN, E. & SAXEN, L. (1974) Transfilter induction of kidney tubules: Correlation with cytoplasmic penetration into Nuclepore filters. J. Embryol. exp. Morph. 31, 667-686.

WATTERSON, R.L., FOWLER, I. & FOWLER, B.J. (1954) The role of the neural tube in the development of the axial skeleton of the chick. Amer. J. Anat. 95, 337-399.

WEISS, L. & NIR, S. (1979) On the mechanism of transfilter induction of kidney tubules. J. Theor. Biol. 78, 11-20.

WEISS, P. (1949) Nature of vertebrate individuality. The problem of cellular differentiation. Proc. First nat. Cancer Conf., pp. 50-60.

WEISS, P. & FITTON-JACKSON, S. (1961) Fine-structural changes associated with lens determination in the avian embryo. Dev. Biol. 3, 532-554.

WESSELLS, N.K. (Ed.) (1977) Tissue Interactions and Development, W.A. Benjamin, Inc., Menlo Park.

Development in Mammals, Vol. 4, M.H. Johnson, editor

INVOLVEMENT OF HORMONES AND GROWTH FACTORS IN THE DEVELOPMENT OF THE SECONDARY PALATE

Robert M. Pratt

Craniofacial Development Section
Laboratory of Developmental Biology and Anomalies
National Institute of Dental Research
National Institutes of Health
Building 30, Room 405
Bethesda, MD 20205 U.S.A.

1. Glucocorticoids and Secondary Palate Development

    1.1 Physiological effects
    1.2 Glucocorticoids and cleft palate
    Glucocorticoid receptors
    Local metabolism of glucocorticoids

2. Role of the H-2 Locus in Secondary Palate Development
3. Mechanism of Action of Glucocorticoids
4. Epidermal Growth Factor and Palatal Development

Development of the mammalian secondary palate constitutes a complex series of developmental events which depend upon the presence of a number of hormones and growth factors. The palatal processes first appear as outgrowths from the maxillary processes and subsequently grow in a vertical position alongside the tongue (Figure 1). The shelves undergo a rapid reorientation to the horizontal position above the tongue which brings the opposing epithelia into contact at the midline. The medial-edge epithelial cells then undergo a programmed cell death with resorption of the basement membrane and cell remnants, and the two shelves are thereby fused into a single structure constituting the secondary palate.

204

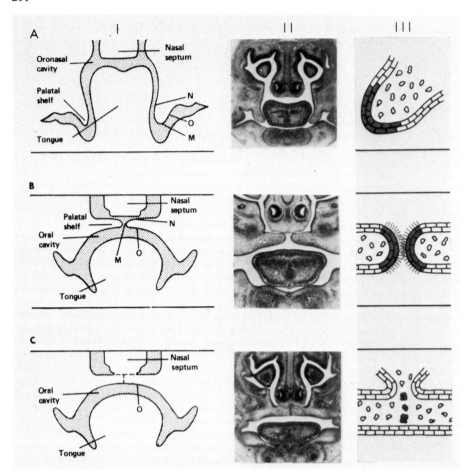

Figure 1: Secondary Palate Development. Schematic frontal section (I) through
the anterior rodent head at the following days of gestation: (A) Day 15-rat or
day 13-mouse; (B) Day 16-rat or day 14-mouse; (C) Day 17-rat or day 15-mouse.
The histologic sections in the middle (II) demonstrate the morphology of Day 14,
15 and 16 rat secondary palates in the anterior region (X 20) (kindly provided
by Dr. Virginia Diewert). The schematics of the palatal shelves to the far
right (III) show that the epithelial cells of the medial-edge (M-shaded area)
cease DNA synthesis (A) and undergo glycoconjugate-mediated (~) adhesion (B).
The palatal shelves eventually fuse and form a continuous structure (secondary
palate) after the degeneration of the medial-edge epithelial cells (C). This
sequence of events does not occur in the epithelial cells of the oral (O) or
nasal (N) epithelium.

Physiological levels of glucocorticoids, in conjunction with
hormones which regulate cyclic nucleotide metabolism, and growth
factors such as epidermal growth factor are essential for normal
growth and differentiation of the epithelial and mesenchymal

cells of the palate, whereas pharmacologic doses of these hormones specifically disrupt palatal development (Salomon & Pratt, 1979). Epidermal growth factor is a polypeptide of approximately 6,000 daltons. It is found in the mouse, in man and undoubtedly in a variety of other mammalian species (Cohen, 1972; Carpenter, 1978; Hollenberg, 1980). In addition to the important biological action of EGF in stimulating cellular proliferation, the occurrence of EGF receptors in a large number of tissues including the placenta (Hollenberg, 1979) and secondary palate (Pratt *et al*, 1978) suggests that EGF may play an important role in embryonic development.

The effects of both glucocorticoids and EGF in the target cells of adult and fetal tissues are mediated through the interaction of the hormone with specific receptor proteins (Figure 2). Following entry, via passive diffusion, of the glucocorticoid into the cell, the steroid binds reversibly to the cytosol receptor and the activated steroid-receptor complex then translocates to the nucleus. Interaction of the hormone-receptor complex with specific sites on the nuclear chromatin results in alterations in transcriptional processes. These molecular events are ultimately expressed as specific cellular responses. In contrast, EGF binds to cell surface glycoprotein receptors and induces aggregation of the hormone receptor complex, internalization of the complex and subsequent lysosomal-mediated degradation. EGF most likely exerts its effect on proliferation and differentiation by the binding to and aggregation of receptors at the cell surface.

1.   GLUCOCORTICOIDS AND SECONDARY PALATE DEVELOPMENT

1.1 *PHYSIOLOGICAL EFFECTS*

Glucocorticoids at physiological levels (10^{-9}M) promote thymidine incorporation and stimulate palatal mesenchyme cell growth *in vitro* (Salomon & Pratt, 1978). Therefore glucocorticoids either alone or through interactions with other hormones or growth factors may function to control certain stages of normal palato-genesis. It has been shown that the growth of primary cultures of mouse and chick embryonic fibroblasts (Cox & MacLeod, 1962; Fodge & Rubin, 1975) and established mouse fibroblast lines such

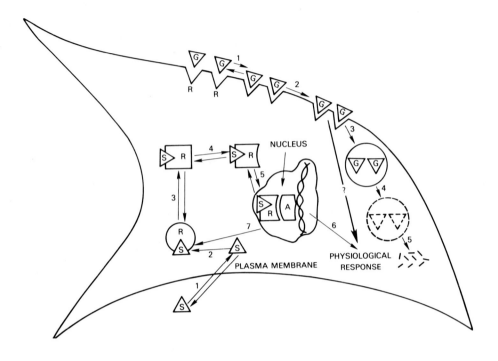

Figure 2: Physiological and molecular mechanisms of action of glucocorticoids and epidermal growth factor. Steroid hormone(s) (S) diffuse across the plasma membrane (1) and bind (2) to specific, high affinity cytoplasmic receptor proteins (R) which in turn undergo a conformational change (3). The steroid-receptor complex subsequently undergoes 'activation' (4), a metabolically dependent process facilitating translocation (5) and binding of the complex to specific nuclear acceptor proteins (A) in chromatin resulting in the modulation of synthesis of specific mRNA molecules and specific physiologic responses (6). In contrast, epidermal growth factor (G) binds (1) to cell surface glyco-protein receptors (R) which is followed by patching (2) of the hormone-receptor complex, internalization of the complex (3), and subsequent lysosomal-mediated degradation (4 and 5). Epidermal growth factor may exert its effect on pro-liferation by some actions at the cell surface.

as L929 and 3T3 (Ruhmann & Berliner, 1965; Pratt & Aronow, 1966; Armelin, 1973; Thrash et al, 1974; Armelin & Armelin, 1977, 1978) is regulated by glucocorticoids, epidermal growth factor (EGF) and fibroblast growth factor (FGF) (Gospodarowicz et al, 1978). Thus, glucocorticoids can either inhibit or stimulate DNA synthesis and cell growth depending upon the origin of the cells and the culture conditions employed in vitro. The stimulation of growth of fibroblast cells *in vitro* is known to be due to a synergistic interaction between glucocorticoids and one of several poly-peptide hormones (Baseman & Hayes, 1975; Armelin, 1973; Gospodarowicz & Moran, 1976; Gospodarowicz *et al,* 1978; Armelin & Armelin, 1978). Current evidence suggests that glucocorticoids may function physiologically by increasing either the number of cell surface receptors for epidermal growth factor and insulin or the affinity of these receptors for their respective ligands (Baker *et al,* 1978).

Glucocorticoids are also known to regulate the production of extracellular matrix components *in vivo* and *in vitro* (Nacht & Garzon, 1974; Moscatelli & Rubin, 1977; Newman & Cutroneo, 1978). The extracellular matrix (collagen, fibronectin, sulfated proteo-glycans and hyaluronic acid) functions as an important determinant during development and differentiation in many tissues of the embryo (see Saxen *et al,* this volume). Components of the matrix may serve to control morphogenetic movements and cell-cell or tissue-tissue interactions such as occur between the epithelium and adjacent mesenchyme in the developing pancreas, thyroid, salivary glands, lung, kidney and mammary gland as well as in the secondary palate (LeDouarin, 1970; Rutter *et al,* 1973; Greene & Pratt, 1976; Tyler & Koch, 1977). In the secondary palate, inter-action between the epithelium and mesenchyme is a necessary requirement for the subsequent autolysis of the medial-edge epithelial cells (Tyler & Koch, 1977).

The palatal shelves synthesize and accumulate sulfated proteo-glycans and hyaluronate prior to and during shelf elevation (Pratt *et al,* 1973). Collagen also rapidly accumulates in the secondary palate during days 11 through 18 of gestation and ultimately accounts for approximately 5% of the total palatal protein *in vivo* (Pratt & King, 1971). Alterations in glycosaminoglycan or collagen metabolism by drugs such as chlorcyclizine, which promotes the degradation of hyaluronic acid (Wilk *et al,* 1978), and β-amino-proprionitrile, which inhibits collagen crosslinking (Pratt & King, 1972) can cause failure of fusion of the palates (cleft palate) by delaying shelf elevation.

208

1.2 GLUCOCORTICOIDS AND CLEFT PALATE

Glucocorticoids, when administered in pharmacological doses (10^{-6} - 10^{-8}M) to rabbits, hamsters, rats, and mice at mid-gestation, inhibit closure of the fetal secondary palate (Baxter & Fraser, 1950; Walker, 1967; Nanda *et al*, 1970; Shah & Kilistoff, 1976). Different strains of inbred mice exhibit different degrees of susceptibility to glucocorticoid-induced cleft palate (Fraser & Fainstat, 1951; Kalter, 1954, 1965). For example, all of the offspring of A/J mice treated with glucocorticoids between days 11 and 14 of gestation have cleft palate, whereas C57BL/6J (C57) mice treated with the same dose of steroid produce only 20-25% off-spring with cleft palate.

A number of mechanisms have been proposed to explain the different responses of various strains of mice to the teratogenic actions of glucocorticoids. Fraser and co-workers have shown that the palatal shelves reach the horizontal position later in normal development in the A/J than in the C57 strain. They suggested that the A/J mice were more susceptible to cortisone, since their palates elevated later making it easier to delay their elevation beyond the point where they still had the capacity to fuse in the horizontal position (Biddle & Fraser, 1977). However, other explanations have been offered and involve postulating strain differences in the metabolism or cellular action of glucocorticoids. There could be strain differences in (i) the maternal levels of endogenous steroids, (ii) the production of more teratogenic metabolites from the administered steroids, (iii) the distribution of the steroid among maternal and fetal tissues, (iv) effects of the steroid on placental function such as inhibition of transfer of essential nutrients including glucose and amino acids, (v) direct effects on serum levels of these nutrients which could indirectly affect fetal growth and differentiation, and (vi) the sensitivity of the embryo's facial structures to the actions of the steroid.

The synthesis as well as metabolism of glucocorticoids does differ among adult mice of various strains (Badr & Spickett, 1965; Doering *et al*, 1972; Lindberg *et al*, 1972; Shire, 1974), and variations in the level of maternal plasma corticosterone could therefore contribute to the strain differences in sensitivity.

However, variations in the endogenous levels of corticosterone between sensitive and resistant strains do not appear to be correlated with the sensitivity to steroids. A/J and C57 mice have no significant differences in the concentration of maternal or fetal corticosterone during midgestation (Salomon et al, 1979).

Glucocorticoids can cross the rodent placenta without being metabolized (Zarrow et al, 1970; Nguyen-Trong-Tuan et al, 1971). When administered in pharmacological doses to pregnant mice, a large fraction of the administered steroid reaches the fetus unmetabolized, and functions as the active teratogenic agent (Zimmerman & Bowen, 1972b). The amount of steroid which reaches the fetus is dependent in part upon the mother's ability to metabolize glucocorticoids (Zimmerman & Bowen, 1972a). Most evidence suggests that glucocorticoid teratogenicity is due to a direct action on the embryo (Zimmerman et al, 1970). The strain dependent effects would therefore need to be explained in terms either of variable ability to bind the steroids (receptor defect) or in terms of variable fetal metabolism of the steroid.

Glucocorticoid receptors

A correlation has been observed between the number of gluco-corticoid receptors in cell lines of mouse lymphosarcoma and fibroblast cells and the growth inhibitory response of these cells to steroids (Yamamoto et al, 1976; Bourgeois & Newby, 1977; Hackney et al, 1970; Venetianer et al, 1978). The total binding of radioactive glucocorticoids to proteins extracted from fetal tissues also varies between different strains of mice and cor-relates with the degree of sensitivity of these strains to gluco-corticoids. For example, a greater amount of bound radioactive steroid administered *in vivo* is associated with fetal A/J tissues than resistant C57 or CBA/J tissues (Levine et al, 1968; Reminga & Avery, 1972). Recent evidence has demonstrated that mouse embryonic facial (maxillary) and palatal mesenchyme cells possess high affinity, specific receptor proteins for glucocorticoids (Salomon & Pratt, 1976, 1978, 1979; Pratt et al, 1979). The synthetic glucocorticoids (dexamethasone and triamcinolone acetonide) have higher affinities for these receptors than the natural glucocorticoid, hydrocortisone (Cake & Litwack, 1975) and their affinities correlate well with their teratogenic potencies

(Pinsky & DiGeorge, 1965; Biddle & Fraser, 1977). While the glucocorticoid receptors in the palatal shelves themselves have not been biochemically characterized, receptors prepared from whole midgestation A/J and C57 mouse embryos have been studied (Salomon et al, 1978). Embryonic cytosols from both strains contain high-affinity, limited capacity, binding proteins for dexamethasone and triamcinolone acetonide. These proteins bind both natural and synthetic glucocorticoids, and can be distinguished from serum corticosteroid binding globulin (transcortin).

A/J facial mesenchyme cells, both freshly isolated and in primary cultures, possess two to three times more cytoplasmic receptors than C57 mesenchyme cells (Salomon & Pratt, 1976). Strain differences in the concentration of cytoplasmic receptors for glucocorticoids in the day 14 embryo are generally restricted to the oral-facial regions (Salomon & Pratt, 1979). Although there are slight differences in the concentration of receptor sites in the forepaw between A/J and C57 mice, these differences are less than those observed in the oral-facial region. The concentration of receptor sites in C57 maxillary and forepaw cells are approximately equivalent, while the concentration of receptor sites in A/J maxillary cells is three to four times higher than in the cells of the forepaw. The presence of glucocorticoid receptors in the fetal forepaw in the mouse is itself interesting since glucocorticoids are known to cause specific limb mal-formations in the mouse (Jurand, 1968).

Studies with strains of mice other than A/J and C57 and also with the rat have shown that the levels of cytoplasmic gluco-corticoid receptors in maxillary mesenchyme cytosols are also correlated with the sensitivity of these animals to glucocorticoid-induced cleft palate (Salomon & Pratt, 1978, 1979; Pratt et al, 1979). For example, maxillary mesenchyme cells obtained from day 14 Swiss Webster (SWR/NIH) embryos have high levels of cyto-plasmic glucocorticoid receptor proteins, and are sensitive to steroid-induced cleft palate. The concentration of steroid receptors in tissues of the oral-facial region is higher than found in the brain or liver (Salomon & Pratt, 1979). The distri-bution of glucocorticoid receptors within the secondary palate is not known, but studies using autoradiography of cryostat sectioned, labelled fetal heads are in progress.

Goldman *et al* (1978) have also presented evidence for the presence of glucocorticoid receptors in cultured mesenchyme cells derived from fused palates of the *human* fetus. Glucocorticoid receptor levels were measured by whole cell binding assays using both dexamethasone and triamcinolone acetonide (Baxter *et al*, 1975). The specific uptake of glucocorticoids into whole cells measures the total steroid receptor concentration in the cell in both the cytoplasmic and nuclear compartments. Androgens or oestrogens did not inhibit dexamethasone uptake in the human mesenchyme cells while glucocorticoids competed with dexamethasone. A variety of structurally unrelated teratogenic (Phenytoin, pheno-barbital and chlorpromazine) and non-teratogenic (Diazepam and barbital) drugs also competed with the binding of dexamethasone in whole cell suspensions. The authors suggested that the teratogenic action of these drugs might be due to competition with dexamethasone for binding directly at the receptor level or to an inhibition of dexamethasone uptake into the cells.

Local metabolism of glucocorticoids

Whilst binding studies suggest that strain differences in susceptibility to glucocorticoids might be due to differences in receptor level or affinity, alternative explanations are possible. For example, it is known that a number of hydrophobic compounds such as polycyclic hydrocarbons and steroids are metabolized by mixed-function oxygenases such as aryl hydrocarbon hydroxylase (Conney, 1967) or the terminal oxidases associated with them (cytochrome P450 and P448) (Gillette, 1966). Certain drugs, polycyclic hydrocarbons and pesticides can inhibit the synthesis as well as enhance the degradation of steroid hormones by affecting the activity of drug metabolizing enzymes (Kupfer, 1975; Kupfer & Bulger, 1976). Since receptor levels were measured by Goldman *et al* (1978) in cell suspensions after 60 minutes at 37°C and not at 0°C with isolated cytosol preparations, these drugs might be enhancing the degradation of dexamethasone by cells in human palatal tissue. As a consequence of this metabolism, less free steroid would be available for receptor binding, which might be interpreted as a direct competition between the drug and the steroid for the receptor.

Glucocorticoid receptors measured in isolated cytosols at 0°C from Swiss Webster embryonic maxillary cells do exhibit specificity for glucocorticoids (Salomon & Pratt, 1978, 1979). In studies under these conditions, drugs such as diphenylhydantoin or phenobarbital, dimethylbenzanthracene and retinoic acid were not found to compete with dexamethasone for binding to specific receptors even in concentrations 1000-fold higher than the steroid. Furthermore, although some of the drugs tested by Goldman (1978) do compete for the specific binding of dexamethasone to glucocorticoid receptors in isolated cytosols from rat hepatoma cells, significant competition is only observed at concentrations which are 50,000 to 200,000-fold excess over the concentration of steroid (Ballard et al, 1975). These concentrations of drugs are well above their plasma therapeutic dose levels. Therefore, it is unlikely that the pharmacological or teratogenic effects of these drugs are mediated through a glucocorticoid receptor. There are cytoplasmic receptors in mouse liver and established cell lines which bind the various drugs and carcinogens (polycyclic hydrocarbons) that induce aryl hydrocarbon hydroxylase activity (Poland & Glover, 1976; Guenthner & Nebert, 1977), but these cytoplasmic receptors are both biochemically and stereospecifically distinct from the steroid hormone receptors and do not bind any steroids.

The sensitivity of different mouse strains to the induction of aryl hydrocarbon hydroxylase by steroids, drugs and polycyclic hydrocarbons is regulated by an autosomal dominant gene, the Ah locus (Nebert et al, 1975). Genetic differences in the inducibility of aryl hydrocarbon hydroxylase are maintained in vitro in fetal mouse fibroblast cell cultures (Nebert & Bausserman, 1970). In fact it has been suggested that the teratogenicity of various drugs in different strains of mice can be predicted by the extent of their metabolism to active epoxides in cell cultures (Nebert, 1973).

In conclusion, whilst strain differences in the levels of cytoplasmic glucocorticoid receptors might partially account for the variations observed in vivo in the differential susceptibility of these strains to the teratogenic effect of glucocorticoids, there may be other differences between these strains, such as the

metabolism or distribution of glucocorticoids in both the
maternal and fetal compartments, which could also contribute to
the etiology of steroid susceptibility.

2. ROLE OF THE H-2 LOCUS IN SECONDARY PALATE DEVELOPMENT

Numerous studies have suggested that the sensitivity of different mouse strains to glucocorticoids is multigenic (Baxter &
Fraser, 1950; Fraser & Fainstat, 1951; Kalter, 1954; Walker &
Fraser, 1957; Loevy, 1963; Kalter, 1965; Dostal & Jelinek, 1973).
Reciprocal crosses of A/J (sensitive) with C57 or CBA/J
(resistant) strains have been studied. The offspring of A/J
mothers have a greater sensitivity to cleft palate after steroid
administration than seen in interstrain crosses using A/J fathers.
Similarly, transfer of A/J blastocysts into CBA/J foster mothers
reduces the cleft palate frequency (Larsson et al, 1977). These
results suggest that the strain sensitivity to glucocorticoids is
determined not only by the fetal genotype but also by maternal
factors. It has been demonstrated that the enhanced sensitivity
of A/J mice compared to C57 mice to steroid-induced cleft palate
is due to two or three independent genetic loci. One of these
loci may be associated with, or linked to, the major histocompatibility locus H-2 in the mouse (Biddle & Fraser, 1977).

The H-2 locus is a gene complex on chromosome 17 (Demant, 1973;
Snell et al, 1973). A variety of genetic loci are in this region
including those genes specifying certain transplantation antigens
(H2-K and H2-D regions), the immune response for both cellular and
humoral reactions (Ir region), a serum protein, (SS-Slp), which
is the fourth component of complement (Meo et al, 1975), and those
gene(s) regulating several androgen- and glucocorticoid-dependent
responses (HOM-1) (Ivanyi et al, 1972a,b; Pla et al, 1976;
Rychlikova & Ivanyi, 1974). Associated with the H-2 locus on
chromosome 17 is the T locus. The T locus interacts with the H-2
locus by suppressing crossing over thereby maintaining polymorphism at the H-2 locus (Snell, 1968). The T locus has an
important role during development of tissues such as the neural
tube (Gluecksohn-Waelsch & Erickson, 1970; Artzt & Bennett, 1975).
Furthermore, expression of the t alleles at the T locus is
influenced both by the H-2 locus and the sex of the animal
(Michova & Ivanyi, 1974).

Allelic genes within the D and K regions of the $H-2$ locus
influence the susceptibility to steroid-induced cleft palate
(Table 1: Bonner & Slavkin, 1975; Tyan & Miller, 1978). Using

TABLE 1:   COMPARISON OF FREQUENCY OF STEROID-INDUCED
CLEFT PALATE IN MICE WITH H-2 HAPLOTYPES

Strain	% Cleft Palate*	H-2 Haplotype
A/J	99	$H-2^a$
B.10A	81	$H-2^a$
SWR/J	100	$H-2^q$
DBA/1J	94	$H-2^q$
C57BL/6J	25	$H-2^b$
C57BL/10 ScSn	21	$H-2^b$
A.By**	85	$H-2^b$
CBA/J	12	$H-2^k$
C3H/HeJ	63	$H-2^k$
bm/bm***	100	$H-2^b$

*   Data obtained from Bonner & Slavkin (1975) and
    Tyan & Miller (1978). Animals administered 2.5mg
    (s.c. or i.m.) cortisone acetate on days 11 through
    14 of gestation.

** High spontaneous cleft lip (greater than A/J),
    (Salomon & Pratt, unpublished observations).

***Brachymorphic mutant (Orkin et al, 1975; Pratt et al,
    1980).

congenic strains of mice, Bonner and Slavkin (1975) demonstrated
that mice of the $H-2^a$ haplotype, A/J and B10.A, were highly susceptible to steroid-induced cleft palate. In contrast, animals
which carried the $H-2^b$ haplotype, C57BL/6J and C57BL/10 ScSn,
were resistant (Table 1). In addition, they showed that the
frequency of cleft palate was higher in hybrid animals bearing a
maternally derived $H-2^a$ haplotype than in hybrids in which the
$H-2^a$ was paternally derived. Likewise, maternal rather than
paternal derivation of the $H-2^b$ haplotype in the hybrid lowered
significantly the frequency of cleft palate production. Tyan and
Miller (1978) have confirmed these results and have further shown

that animals bearing the H-2^q alleles (SWR/J and DBA/1J), like those possessing the H-2^a, are sensitive to steroid-induced cleft palate.

The H-2 locus is not the sole determinant of cleft palate sensitivity since C3H/HeJ and CBA/J mice have the same H-2 haplotype $(H$-$2^k)$ but different susceptibilities to steroid-induced cleft palate (Goldman et al, 1977) (Table 1). Similarly, the C57 and bm/bm mice have the same H-2 haplotype but quite different frequencies of cleft palate (Table 1), and both A/J and B10.A strains have the same H-2 haplotype $(H$-$2^a)$, but the frequency of cleft palate production by glucocorticoids in these strains is not identical (Bonner & Slavkin, 1975). Therefore, additional non-H-2 associated genetic differences between these strains must also contribute to the in vivo response to glucocorticoids. Since there may be differences in genes other than those in the H-2 region in congenic H-2 strains, genes which are tightly linked to H-2 such as the T locus (Snell, 1968) might influence the response of these strains to glucocorticoids.

The mechanism(s) by which the strain differences in gluco-corticoid susceptibility is determined by the H-2 locus or other genes is not clear. These differences could be due to the control by the locus of steroid receptor levels in the palate. Goldman et al (1977) have demonstrated that the concentration of hydro-cortisone receptors is higher in the secondary palates in animals possessing the H-2^a haplotype than in those having the H-2^b haplotype. While this association between glucocorticoid receptor levels and the H-2 haplotype exists in the palate, Butley et al (1978) were unable to find a similar correlation in adult mouse liver. They demonstrated that A/J hepatic tissue contains twice as many cytoplasmic glucocorticoid receptors as C57 hepatic tissue. However, A/J $(H$-$2^a)$ and A.By $(H$-$2^b)$ although possessing different H-2 haplotypes had comparable hepatic levels of gluco-corticoid receptors. Likewise, although C57 $(H$-$2^b)$ and B10.A $(H$-$2^a)$ strains had lower concentrations of receptor sites than A/J and A.By, the amount of receptors in C57 and B10.A livers was equivalent. These results would indicate that H-2 or other genes closely associated with this locus only influence receptor levels in the secondary palate.

3. MECHANISM OF ACTION OF GLUCOCORTICOIDS

Glucocorticoids (at 10^{-6} to 10^{-8}) affect primary cultures of mouse embryonic palate mesenchyme cells *in vitro* (Salomon & Pratt, 1978), causing a reduction in cell number and a simultaneous decrease in the incorporation of ^{3}H-thymidine into DNA in both A/J and C57 mesenchyme cells. No significant changes in total protein synthesis were observed in either cell type following a 48 hour exposure to steroids. In both A/J and C57 cultures the reduction in thymidine incorporation precedes the steroid-inhibitory effect on cell growth. The decrease in cell growth exhibits a dose-dependent response to dexamethasone and is specific for steroids possessing glucocorticoid activity. Cells such as A/J possessing a high level of glucocorticoid receptors are more sensitive to inhibition of growth and thymidine incorporation than C57 cells which have a lower level of glucocorticoid receptors.

A differential effect on ^{3}H-thymidine incorporation and on growth can also be observed in cultures of palatal shelves from A/J and C57 mice which have been maintained in the presence of dexamethasone or hydrocortisone (Saxen, 1973; Pratt *et al*, 1979). Under these conditions there is a greater inhibition of thymidine incorporation into DNA in A/J secondary palates than in C57 palates. The inhibition is more pronounced in the palatal mesenchyme cells than in the nasal, oral or medial-edge epithelial cell layers. These results are in accord with the observations that dexamethasone and cortisone inhibit *in vivo* the proliferation of mesenchymal cells of the palatal processes in rats and mice (Nanda & Romeo, 1978; Mott *et al*, 1969; Jelinek & Dostal, 1975). The production of cleft palate may therefore arise because the reduced size of the palatal shelves due to reduced cellular proliferation causes failure of the shelves to make adequate contact and fuse at the appropriate developmental stage. During the terminal stages of palate development (Figure 1), there is a transient increase in the endogenous level of palatal cyclic AMP (Figure 3) (Pratt & Martin, 1975; Greene & Pratt, 1979; Pratt *et al*, 1979). This increase precedes or is concomitant with the sequence of biochemical events which are involved in the 'programmed' autolysis of the medial-edge epithelial cells. These

events in the medial-edge epithelium include cessation of DNA
synthesis, acquisition of a glycoconjugate-rich surface coat,
release of lysosomal enzymes and subsequent shelf fusion (Greene &
Pratt, 1976). Recent studies (Greene & Pratt, 1979; Greene et al,
1980) have shown that both adenyl cyclase activity and cyclic AMP
levels in the medial-edge cells increase dramatically and specifically at the later stages of terminal cell differentiation.

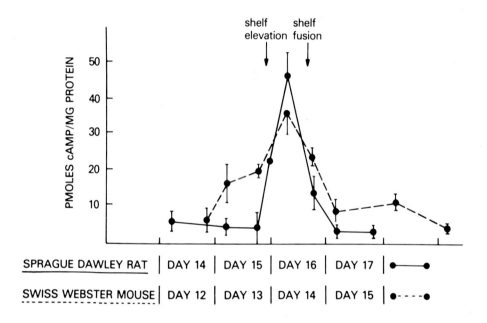

Figure 3: Cyclic AMP levels during rodent secondary palatal development. The
levels of cyclic AMP in whole palatal shelves (mesenchyme plus epithelium) were
determined in the Swiss Webster mouse and Sprague-Dawley rat at various times
during gestation (see Greene & Pratt, 1979).

This increase in adenyl cyclase activity in the medial-edge
epithelial cells was blocked by DON which prevents programmed cell
death in vitro. It still remains to be determined whether this
increase in adenyl cyclase and cyclic AMP serves to influence
terminal cell differentiation or just reflects programmed degenerative changes in these cells.

The genetic factors which control the response of the medial-edge epithelial cells to agents that influence the intracellular levels of cyclic AMP are not known. Although Meruelo and Edidin (1975) have demonstrated an association between the *H-2* haplotype and the endogenous levels of cyclic AMP in livers of adult *H-2* congenic strains of mice, Erickson and Butley (1977) were unable to find a similar correlation in the mouse secondary palate. The endogenous levels of cyclic AMP were higher in the palatal shelves of A/J *(H-2^a)* and A.By *(H-2^b)* strains than the endogenous levels of cyclic AMP in C57BL/10J *(H-2^b)* and B10.A *(H-2^a)* mice. However, there was no significant difference in cyclic AMP levels in the palate between the two parental strains and their congenic pairs. The mutant brachymorphic mouse is extremely sensitive to cortisone-induced cleft palate and a recent study (Pratt *et al*, 1980) demonstrates that elevated levels of palatal cyclic AMP are found in the brachymorphic genotype compared to the wild type.

Although the endocrine factors which cause an increase in the levels of cyclic AMP in the palate have not been fully characterized, glucocorticoids in conjunction with epidermal growth factor may be involved. Epidermal growth factor inhibits the normal degeneration of the medial-edge epithelial cells *in vitro*, and can promote hypertrophy and keratinization of these cells (see next section and Hassell, 1975; Hassell & Pratt, 1977; Pratt *et al*, 1979). The response to epidermal growth factor can be prevented by the addition of dibutyryl cyclic AMP (Hassell & Pratt, 1977). Since the levels of cyclic AMP are generally elevated in non-growing cells (Friedman, 1976), and since glucocorticoids can inhibit the growth of a variety of cell types (Cox & MacLeod, 1962; Nacht & Gagon, 1974), these steroids might also function to influence the levels of cyclic AMP in the medial-edge epithelial cells and the subsequent mitotic activity of these cells.

Glucocorticoids also accelerate the keratinization of embryonic epidermal cells (Sugimoto *et al*, 1976) and modify the extracellular composition of the cell surface by regulating components of the cytoskeleton (Berliner & Gerschenson, 1975) and by producing alterations in the composition of surface glycoproteins (Behrens & Hollander, 1976). These changes are probably related

to the increase in adhesive properties of cells treated with
glucocorticoids (Ballard & Tomkins, 1970; Lotem & Sachs, 1975).
Therefore, glucocorticoids like epidermal growth factor may be
acting directly on the medial-edge epithelial cells to influence
adhesion between the opposing palatal shelves.

The *H-2* complex may also have a role in controlling the
appearance of the glycoconjugate-rich surface coat on the medial-
edge epithelial cell surface prior to contact and adhesion of the
opposing palatal shelves. Acquisition of the glycoprotein coat
is necessary for the adhesion and subsequent fusion of the palatal
shelves *in vivo* and *in vitro* (Pratt *et al*, 1973; Pratt & Hassell,
1975; Greene & Pratt, 1976). Cell surface glycoproteins are
known mediators of cell-cell recognition during embryonic growth
and differentiation (Moscona, 1974; Edelman, 1976). Moreover,
surface glycoprotein antigens of the *H-2* complex determine the
specificity and rate of intercellular adhesion between embryonic
but not adult fibroblasts *in vitro* (Bartlett & Edidin, 1978).
Therefore, the *H-2* locus may regulate the expression of the
glycoprotein coat on the medial-edge epithelial cells and control
the specificity of adhesion between the homologous shelves in
contrast to adhesion with other regions of epithelium in the oral
or nasal cavity.

Shapira and Shoshan (1972) have demonstrated that the *in vivo*
administration of pharmacological levels of cortisone to pregnant
mice causes a reduced synthesis of collagen in fetal palates.
This effect can be duplicated *in vitro* in maxillary mesenchyme
cells obtained from A/J or C57 embryos. Both A/J and C57 cultured
mesenchyme cells synthesize collagen which constitutes approximately 7 to 10% of the total protein. Cells cultured in the
presence of dexamethasone (10^{-6}M) for six days exhibit a decrease
in total protein synthesis of both collagen and non-collagen proteins. There is a greater reduction in collagen synthesis in A/J
cells than in comparably treated C57 cells (Salomon & Pratt, 1979).

Collagen may be one factor which regulates the growth of non-
transformed cells (Liotta *et al*, 1978). Arrest of collagen
synthesis or secretion by a specific proline analog, cis-
hydroxyproline, produces a cessation of cell growth; providing
treated cells with a collagen substrate can restore normal cell

proliferation. Therefore, the control of collagen synthesis by
hormones may be important for both cell growth and differentiation.
Failure of either the palatal mesenchyme or epithelial cells to
synthesize specific matrix components or cell surface glycoproteins which are necessary for normal mesenchyme-epithelial
interactions could result in abnormal "programming" of the medial-
edge epithelial cells or abnormal growth of these two cell types.

4. EGF AND PALATAL DEVELOPMENT

Organ culture studies have demonstrated that cell death
within the medial palatal epithelium does not depend upon contact
with an apposing palatal process (Smiley & Koch, 1972; Tyler &
Koch, 1975), nor does it depend upon the presence of an underlying
mesenchyme for at least three days prior to fusion (Tyler & Koch,
1977). Cell death occurs within the medial region of cultured
palatal epithelium that has been isolated from its mesenchyme,
and the timing of this cell death proceeds according to the *in
vivo* schedule. Other *in vitro* studies on intact palatal processes
have shown that the cessation of DNA synthesis and cell death
within the medial palatal epithelium can be inhibited in organ
culture by the addition of EGF to the culture medium (Hassell,
1975; Pratt *et al*, 1978). If administered after the cessation of
DNA synthesis within the medial epithelium, EGF does not reinitiate
DNA synthesis, but does inhibit cell death within this region
(Hassell & Pratt, 1977).

In contrast, if present in the culture medium of shelves
explanted prior to cessation of DNA synthesis (mouse, day 12 of
gestation), EGF prevents both the cessation of DNA synthesis and
cell death within the medial epithelium (Figure 4). Tyler and
Pratt (1980) have recently shown that EGF does not stimulate DNA
synthesis or prevent medial epithelial cell death in cultured
isolated palatal epithelium taken at day 13 or 14 of gestation.
EGF did however affect the morphology of the isolated epithelium.
These results suggest that EGF does affect the cells of the
isolated palatal epithelium and that these effects are independent
of the effect of EGF on cell proliferation in the palatal
epithelium. The results of this study showed further that EGF
does stimulate DNA synthesis in palatal epithelium cultured trans-

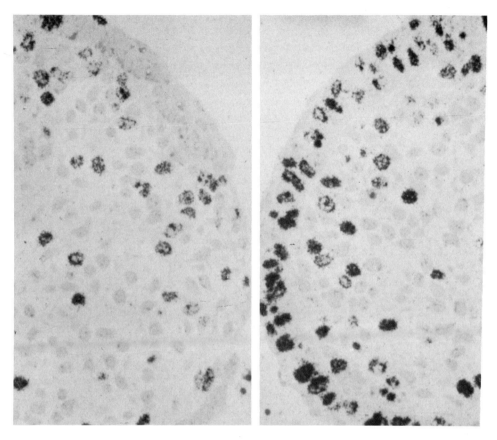

Figure 4: Effect of EGF on cultured mouse secondary palates. Swiss Webster mouse palates were explanted on day 12 into organ culture for 72 hours in Ham's F12 medium containing 10% fetal calf serum plus antibiotics. These are homologous shelves from the same embryo cultured in the absence (A) or presence (B) of EGF at 20 ng/ml for 72 hours with medium change every 24 hours. The cultures were exposed to ^{3}H-thymidine at 1μCi/ml from 66-72 hours of incubation, fixed in glutaraldehyde, embedded in Spurr plastic and processed for autoradiography. Mag. X 930.

filter to the palatal mesenchyme. This result indicates that the palatal epithelium requires the presence of the mesenchyme in order to respond to the mitogenic effects of EGF. It is most likely that EGF is acting directly upon the epithelium and that the mesenchyme provides an extracellular matrix substrate, composed primarily of collagen and fibronectin, that permits the epithelium to respond to EGF.

Recent findings demonstrate that EGF receptors appear in the
embryonic mouse on the eleventh to twelfth day of gestation and
can be detected in embryos (Figure 5) in the palatal epithelium
on day 13 of gestation by autoradiographic localization (Figure 6)
(Nexo et al, 1980). Furthermore, this study also demonstrated
the presence of an EGF-like substance in mouse embryos during

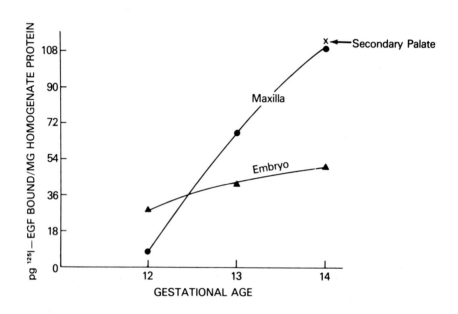

Figure 5: Binding of ^{125}I-EGF by embryonic mouse homogenates. The binding of
^{125}I-EGF to membranes from homogenates of maxilla (●), secondary palate (x)
and whole embryo (▲) was measured according to Nexo et al, 1980.

these stages of development. The source and exact composition of
the EGF extracted from the embryos is uncertain although maternal
EGF could contribute to fetal levels via transplacental transport.
This is unlikely since maternal and fetal EGF differ in that the
fetal form is not extremely reactive, as is maternal EGF, with
anti-EGF antibodies, but is reactive with EGF receptors. This
kind of reactivity is reminiscent of the sarcoma growth factor
that can be isolated from murine or feline sarcoma virus-infected
mouse cells (Delarco & Todaro, 1978). These results suggest that

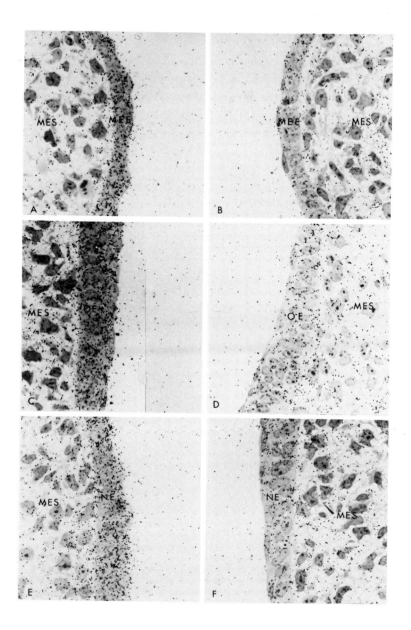

Figure 6: Autoradiographic localization of ^{125}I-EGF binding in the mouse
secondary palate. Day 13 mouse shelves were dissected and incubated in phos-
phate buffered saline (PBS) containing ^{125}I-EGF according to Nexo *et al*, 1980.
A, C & E (left side) - ^{125}I-EGF only; B, D & F (right side) - ^{125}I-EGF plus
1µg/ml unlabeled EGF; MEE - Medial-edge epithelium (A & B); OE - Oral
epithelium (C & D); NE - Nasal epithelium (E & F), MES - mesenchyme.

the fetal form of mouse EGF may differ from the molecule present in the adult and is likely to be of fetal origin.

The results of several studies suggest that the palate might be dependent upon EGF for some aspect of its growth and differentiation. Recent studies (Pratt, unpublished) demonstrated that palatal shelves can be successfully cultured in serum-free medium containing EGF among other growth factors and hormones. The selective removal of EGF from the medium results in a dramatically reduced overall growth and in the death of the palatal medial epithelial cells. Therefore, EGF may be important for the growth and differentiation of both palatal mesenchyme and epithelial cells.

REFERENCES

ARMELIN, H.A. (1973) Pituitary extracts and steroid hormones in the control of 3T3 cell growth. Proc. Natl. Acad. Sci. USA 70, 2702-2706.

ARMELIN, M.C.S. & ARMELIN, H.A. (1977) Serum and hormonal regulations of the "resting-proliferative" transition in a variant of 3T3 mouse cells. Nature 265, 148-151.

ARMELIN, M.C.S. & ARMELIN, H.A. (1978) Steroid hormones mediate reversible phenotypic transition between transformed and untransformed states in mouse fibroblasts. Proc. Natl. Acad. Sci. USA 75, 2805-2809.

ARTZT, K. & BENNETT, D. (1975) Analogies between embryonic (T/t) antigens and adult major histocompatibility (H-2) antigens. Nature 256, 545-547.

ASUA, L. J-d., CARR, B., CLINGAN, D. & RUDLAND, P.S. (1977) Specific glucocorticoid inhibition of growth promoting effects of prostaglandin F2α on 3T3 cells. Nature 265, 450-452.

ASUA, L. J-d., CLINGAN, D. & RUDLAND, P.S. (1975) Initiation of cell proliferation in cultured mouse fibroblasts by prostaglandin F2α. Proc. Natl. Acad. Sci. USA 72, 2724-2728.

BADR, F.M. & SPICKETT, S.G. (1965) Genetic variation in the biosynthesis of corticosteroids in Mus musculus. Nature 205, 1088-1090.

BAKER, J.B., BARSH, G.S., CARNEY, D.H. & CUNNINGHAM, D.C. (1978) Dexamethasone modulates binding and action of epidermal growth factor in serum-free cell culture. Proc. Natl. Acad. Sci. USA 75, 1882-1886.

BALLARD, P.L., CARTER, J.P., GRAHAM, B.S. & BAXTER, J.D. (1975) A radio-receptor assay for evaluation of the plasma glucocorticoid activity of natural and synthetic steroids in man. J. Clin. Endocrinol. Metab. 41, 290-304.

BALLARD, P.L. & TOMKINS, G.M. (1970) Glucocorticoid-induced alteration of the surface membrane or cultured hepatoma cells. J. Cell Biol. 47, 222-234.

BARTLETT, P.F. & EDIDIN, M.L. (1978) Effect of the H-2 gene complex on rates of fibroblast intercellular adhesion. J. Cell Biol. 77, 377-388.

BASEMAN, J.B. & HAYES, N.S. (1975) Differential effect of hormones on macro-molecular synthesis and mitosis in chick embryo cells. J. Cell Biol. 67, 492-497.

BAXTER, J.D., HIGGINS, S.J. & ROUSSEAU, G.G. (1975) Measurement of specific binding of a ligand in intact cells: Dexamethasone binding by cultured hepatoma cells. In: B.W. O'Malley and J.G. Hardman (Eds.), Methods in Enzymology, Vol.XXXVI, Academic Press, New York, pp. 240-248.

BAXTER, J.D. & TOMKINS, G.M. (1971) Specific cytoplasmic glucocorticoid hormone receptors in hepatoma tissue culture cells. Proc. Natl. Acad. Sci. USA 68, 932-937.

BEHRENS, U. & HOLLANDER, V.P. (1976) Cell membrane sialoglycopeptides of corticoid-sensitive and -resistant lymphosarcoma P1798. Caner Res. 36, 172-180.

BERLINER, J.A. & GERSHENSON, L.E. (1975) The effects of a glucocorticoid on the cell surface of RLC-GAI cells. J. Cell Physiol. 86, 523-532.

BIDDLE, F.G. & FRASER, F.C. (1977) Cortisone-induced cleft palate in the mouse: A search for the genetic control of the embryonic response trait. Genetics 85, 289-302.

BONNER, J.J. & SLAVKIN, H.C. (1975) Cleft palate susceptibility linked to histocompatibility-2(H-2) in the mouse. Immunogenetics 2, 213-218.

BOURGEOIS, S. & NEWBY, R.F. (1977) Diploid and haploid states of the glucocorticoid receptor gene of mouse lymphoid cell lines. Cell 11, 423-430.

BUTLEY, M.S., ERICKSON, R.P. & PRATT, W.B. (1978) Hepatic glucocorticoid receptors and the H-2 locus. Nature 275, 136-138.

CARPENTER, G. (1978) The regulation of cell proliferation: Advances in the biology and mechanism of action of EGF. J. Invest. Dermat. 71, 283-287.

COHEN, S. (1962) Isolation of a mouse submaxillary gland protein acceleration incisor eruption and eyelid opening in the newborn animal. J. Biol. Chem. 237, 1555-1562.

CONNEY, A.H. (1967) Pharmacological implications of microsomal enzyme induction. Pharmacol. Rev. 19, 317-366.

COX, R.P. & MacLEOD, C.M. (1962) Alkaline phosphatase content and the effects of prednisolone on mammalian cells in culture. J. Gen. Physiol. 45, 439-485.

DEMANT, P. (1973) H-2 gene complex and its role in alloimmune reactions. Transplant. Rev. 15, 162-200.

DOERING, C.H., SHIRE, J.G.M., KESSLER, S. & CLAYTON, R.B. (1972) Cholesterol ester concentrations and corticosterone production in adrenals of the C57BL/10 and DBA/2 strains in relation to adrenal lipid depletion. Endocrinology 90,

DOSTAL, M. & JELINEK, R. (1973) Sensitivity of embryos and interspecies differences in mice in response to prenatal administration of corticoids. Teratology 8, 245-252.

EDELMAN, G.M. (1976) Surface modulation in cell recognition and cell growth. Science 192, 218-226.

ERICKSON, R.P. & BUTLEY, M.S. (1977) Biochemical basis for H-2 associated sensitivity to steroid-induced cleft palate in mice. Abst. Amer. Soc. Human Gen. 40A.

226

FODGE, D.W. & RUBIN, H. (1975) Differential effects of glucocorticoids on
DNA synthesis in normal and virus-transformed chick embryo cells. Nature
257, 804-806.

FRASER, F.C. & FAINSTAT, T. (1951) Product of congenital defects with
particular reference to cleft palate. Ped. Springfield 8, 527-533.

FRIEDMAN, D.L. (1976) Role of cyclic nucleotides in cell growth and
differentiation. Physiolog. Rev. 56, 652-708.

GILLETE, J.R. (1966) Biochemistry of drug oxidation and reduction by enzymes
in hepatic endoplasmic reticulum. Adv. Pharmacol. 4, 219-261.

GLUECKSOHN-WAELSCH, S. & ERICKSON, R.P. (1970) The T-locus of the mouse:
Implications for mechanisms of development. In: A.A. Moscona and A. Monroy
(Eds.), Current Topics in Developmental Biology, Vol. 5, Academic Press,
New York, pp. 281-316.

GOLDMAN, A.S., KATSUMATA, M., YAFFE, S.J. & GASSER, D.L. (1977) Palatal
cytosol cortisol-binding protein associated with cleft palate susceptibility and H-2 genotype. Nature 265, 643-645.

GOLDMAN, A.S., SHAPIRO, B.H. & KATSUMATA, M. (1978) Human foetal palatal
corticoid receptors and teratogens for cleft palate. Nature 272, 464-466.

GOSPODAROWICZ, D., GREENBERG, G., BIALECHKI, H. & ZEHER, B.R. (1978) Factors
involved in the modulation of cell proliferation in vivo and in vitro: The
role of fibroblast and epidermal growth factors in the proliferative response of mammalian cells. In Vitro 14, 85-118.

GOSPODAROWICZ, D. & MORAN, J.S. (1976) Growth factors in mammalian cell
culture. Ann. Rev. Biochem. 45, 531-558.

GREEN, H. & GOLDBERG, B. (1965) Synthesis of collagen by mammalian cell
lines of fibroblastic and nonfibroblastic origin. Proc. Natl. Acad. Sci.
USA 53, 1360-1365.

GREENE, R.M. & KOCHHAR, D.M. (1975) Some aspects of corticosteroid-induced
cleft palate: A review. Teratology 11, 47-56.

GREENE, R.M. & PRATT, R.M. (1976) Developmental aspects of secondary palate
formation. J. Embryol. exp. Morph. 36, 225-245.

GREENE, R.M. & PRATT, R.M. (1979) Correlation between cyclic-AMP levels and
cytochemical localization of adenylate cyclase during development of the
secondary palate. J. Histochem. Cytochem. 27, 924-931.

GREENE, R.M., SHANFELD, J.L., DAVIDOVITCH, Z. & PRATT, R.M. Exp. Embryo and
Morphol. in press.

GUENTHNER, T.M. & NEBERT, D.W. (1977) Cytosolic receptor for aryl hydrocarbon
hydroxylase induction by polycyclic aromatic compounds. J. Biol. Chem. 252,
8981-8989.

HACKNEY, J.F., GROSS, S.R., ARONOW, L. & PRATT, W.B. (1970) Specific
glucocorticoid-binding macromolecules from mouse fibroblasts growing in vitro.
Mol. Pharmacol. 6, 500-512.

HARVEY, W., GRAHAME, R. & PANAYI, G.S. (1976) Effects of steroid hormones
on human fibroblasts in vitro. II. Antagonism by androgens of cortisol-
induced inhibition. Ann. Rheum. Dis. 35, 148-151.

HASSELL, J.R. (1975) The development of the rat palatal shelves in vitro.
An ultrastructural analysis of the inhibition of epithelial cell death and
palatal fusion by EGF. Develop. Biol. 45(1), 90-102.

HASSELL, J.R. & PRATT, R.M. (1977) Elevated levels of cAMP alters the effect of epidermal growth factor in vitro on programmed cell death in the secondary palate epithelium. Exp. Cell Res. 106, 55-62.

HOLLENBERG, M.D. Vitamins and Hormones, in press.

IVANYI, P., GREGOROVA, S. & MICKOVA, M. (1972b) Genetic differences in thymus, lymph nodes, testes and vesicular gland weights among inbred mouse strains: Association with the major histocompatability (H-2) system. Fol. Biol. (Praha) 18, 81-91.

IVANYI, P., HARMPL, R., STARKA, L. & MICKOVA, M. (1972a) Genetic association between H-2 gene and testosterone metabolism in mice. Nature New Biol. 238, 280-281.

JELINEK, R. & DOSTAL, M. (1975) Inhibitory effect of corticoids on the proliferative pattern in mouse palatal processes. Teratology 11, 193-198.

KALTER, H. (1954) Inheritance of susceptibility to the teratogenic action of cortisone in mice. Genetics 39, 185-196.

KALTER, H. (1965) Interplay of intrinsic and extrinsic factors. In: J.G. Wilson and J. Warkany (Eds.), Teratology: Principles and Techniques, University of Chicago Press, Chicago, pp. 57-64.

KUPFER, D. (1975) Effects of pesticides and related compounds on steroid metabolism and function. Crit. Rev. Toxicol. 4, 83-124.

KUPFER, D. & BULGER, W.H. (1976) Interaction of chlorinated hydrocarbons with steroid hormones. Fed. Proc. 35, 2603-2608.

LARSSON, K.S., MARSK, L., SVONBERG-LARSSON, A. & TESH, J.M. (1977) Agent determined type of maternal influence on induced cleft palate in mice. Biol. Neonate. 31, 51-59.

LeDOURIN, N. (1970) Induction of determination and induction of differentiation during development of the liver and certain organs of endomesodermal origin. In: E. Wolff (Ed.), Tissue Interactions During Organogenesis, Gordon and Breach Science Publ., New York, pp. 37-70.

LEVINE, A.I., YAFFE, S.J. & BACK, N. (1968) Maternal-fetal distribution of radioactive cortisol and its correlation with teratogenic effect. Proc. Soc. Exp. Biol. Med. 129, 86-88.

LINDBERG, M., SHIRE, J.G.M., DOERING, C.H., KESSLER, S. & CLAYTON, R.B. (1972) Reductive metabolism of corticosterone in mice: Differences in NADPH requirements of liver homogenates of males of two inbred strains. Endocrinology 90, 81-92.

LIOTTA, L.A., VEMBU, D., KLEINMAN, H.K., MARTIN, G.R. & BOONE, C. (1978) Collagen required for proliferation of cultured connective tissue cells but not their transformed counterparts. Nature 272, 622-624.

LOEVY, H. (1962) Genetic influences on induced cleft palate in different strains of mice. Anat. Rec. 145, 117-122.

LOTEM, J. & SACHS, L. (1975) Induction of specific changes in the surface membrane of myeloid leukemic cells by steroid hormones. Internatl. J. Cancer 15, 731-740.

LOWRY, O.H., ROSEBROUGH, N.J., FARR, A.L. & RANDALL, R.J. (1951) Protein measurement with folin phenol reagent. J. Biol. Chem. 193, 265-275.

228

MEO, T., KRASTEFF, T. & SHREFFLER, D.C. (1975) Immunochemical
characterization of murine H-2 controlled Ss (serum substance protein)
through identification of its human homologue as the fourth component of
complement. Proc. Natl. Acad. Sci. USA 72, 4536-4540.

MERUELO, D. & EDIDIN, M. (1975) Association of mouse liver adenosine
3':5'-cyclic monophosphate (cyclic AMP) levels with histocompatability-2
genotype. Proc. Natl. Acad. Sci. USA 72, 2644-2648.

MICKOVA, M. & IVANYI, P. (1974) Sex-independent and H-2 linked influence on
expressivity of the Brachyury gene in mice. J. Hered. 65, 369-372.

MONROE, C.B. (1968) Induction of tryptophan oxygenase and tyrosine amino-
transferase in mice. Amer. J. Physiol. 214, 1410-1414.

MOSCATELLI, D. & RUBIN, H. (1977) Hormonal control of hyaluronic acid
production in fibroblasts and its relation to nucleic acid and protein
synthesis. J. Cell. Physiol. 91, 79-88.

MOSCONA, A.A. (1974) Surface specification of embryonic cells: Lectin
receptors, cell recognition, and specific cell ligands. In: A.A. Moscona
(Ed.), The Cell Surface in Development, John Wiley & Sons, New York,
pp. 67-101.

MOSCONA, A.A. (1975) Hydrocortisone-mediated regulation of gene expression
in embryonic neural retina: Induction of glutamine synthetase. J. Steroid
Biochem. 6, 633-638.

MOTT, W.J., TOTO, P.D. & HILGERS, D.C. (1969) Labelling index and cellular
density in palatine shelves of cleft palate mice. J. Dent. Res. 48, 263-265.

NACHT, S. & GARZON, P. (1974) Effects of corticosteroids on connective
tissue and fibroblasts. In: M.H. Briggs and C.A. Christie (Eds.), Advances
in Steroid Biochemistry and Pharmacology, Vol. 4, Academic Press, New York,
pp. 157-187.

NANDA, R. & ROMEO, D. (1978) The effect of dexamethasone and hypervitaminosis A on the cell proliferation of rat palatal processes. Cleft
Palate J. 15, 176-181.

NANDA, R., VAND DER LINDEN, F.P.G.M. & JANSEN, H.W.B. (1970) Production
of cleft palate with dexamethasone and hypervitaminosis A in rat embryos.
Experientia 26, 1111-1112.

NEBERT, D.W. (1973) Use of fetal cell culture as an experimental system
for predicting metabolism in the intact animal. Clin. Pharmacol. and
Therap. 14, 693-704.

NEBERT, D.W. & BAUSSERMAN, L.L. (1970) Genetic differences in the extent
of aryl hydrocarbon hydroxylase induction in mouse fetal cell cultures.
J. Biol. Chem. 245, 6373-6382.

NEBERT, D.W., ROBINSON, J.R., NIWA, A., KUMAKI, K. & POLAND, A.P. (1975)
Genetic expression of aryl hydrocarbon hydroxylase activity in the mouse.
J. Cell Physiol. 85, 393-415.

NEWMAN, R.A. & CUTRONEO, K.R. (1977) Glucocorticoids selectivity decrease
the synthesis of hydroxylated collagen peptides. Mol. Pharmacol. 14, 189-198.

NEXO, E., HOLLENBERG, M.D., FIGUEROA, A. & PRATT, R.M. Proc. Natl. Acad. Sci.
USA, in press.

ORKIN, R.W., PRATT, R.M. & MARTIN, G.R. (1976) Undersulfated glycosaminoglycans in the cartilage matrix of brachymorphic mice. Develop. Biol. 50,
82-94.

PETERKOFSKY, B. & DIEGELMANN, R. (1971) Use of a mixture of proteinase-free collagenases for the specific assay of collagen in the presence of other proteins. Biochemistry 10, 988-993.

PINSKY, L. & DiGEORGE, A.M. (1965) Cleft palate in the mouse: A teratogenic index of glucocorticoid potency. Science 147, 402-403.

PLA, M., ZAKANY, J. & FRACHET, J. (1976) H-2 influence on corticosteroid effects on thymus cells. Folia. Biol. (Praha) 22, 49-50.

POLAND, A., GLOVER, E. & KENDE, A.S. (1976) Stereospecific, high affinity binding of 2,3,7,8-tetrachlorodibenzo-p-dioxin by hepatic cytosol. J. Biol. Chem. 251, 4936-4946.

PRATT, W.B. & ARONOW, L. (1966) The effect of glucocorticoids on protein and nucleic acid synthesis in mouse fibroblasts growing in vitro. J. Biol. Chem. 244, 5244-5250.

PRATT, R.M., FIGUEROA, A.A., NEXO, E. & HOLLENBERG, M.D. (1978) Involvement of EGF during secondary palatal development. J. Cell Biol. 79, 24A.

PRATT, R.M., FIGUEROA, A.A., GREENE, R.M. & SALOMON, D.S. (1979) In: T.V.N. Persaud (Ed.) Abnormal Embryogenesis: Cellular and Molecular Aspects, MTP Press Ltd., Lancaster, pp. 161-176.

PRATT, R.M., GOGGINS, J.R., WILK, A.L. & KING, C.T.G. (1973) Acid mucopoly-saccharide synthesis in the secondary palate of the developing rat at the time of rotation and fusion. Develop. Biol. 32, 230-237.

PRATT, R.M. & HASSELL, J.R. (1975) Appearance and distribution of carbo-hydrate-rich macromolecules on the epithelial surface of the developing rat palatal shelf. Develop. Biol. 45, 192-198.

PRATT, R.M. & KING, C.T.G. (1971) Collagen synthesis in the secondary palate of the developing rat. Arch. Oral Biol. 16, 1181-1185.

PRATT, R.M. & KING, C.T.G. (1972) Inhibition of collagen cross-linking associated with β-aminoproprionitrile-induced cleft palate in the rat. Develop. Biol. 27, 322-328.

PRATT, R.M. & MARTIN, G.R. (1975) Epithelial cell death and cyclic AMP increase during palatal development. Proc. Natl. Acad. Sci. USA 72, 874-877.

PRATT, R.M., SALOMON, D.S., DIEWERT, V.M., ERICKSON, R.P., BURNS, R. & BROWN, K.S. Teratogenesis, Mutagenesis and Carcinogenesis, in press.

REMINGA, T.A. & AVERY, J.K. (1972) Differential binding of ^{14}C-cortisone in fetal, placental, and maternal liver tissue in A/Jax and C57BL mice. J. Dent. Res. 51, 1426-1430.

ROUSSEAU, G.G., BAXTER, J.D. & TOMKINS, G.M. (1972) Glucocorticoid receptors: Relations between steroid binding and biological effects. J. Mol. Biol. 67, 99-115.

RUBIN, H. (1977) Antagonistic effects of insulin and cortisol on coordinate control of metabolism and growth in cultured fibroblasts. J. Cell Physiol. 91, 249-260.

RUHMANN, A.G. & BERLINER, D.L. (1965) Effect of steroids on growth of mouse fibroblasts in vitro. Endocrinology 76, 916-927.

RYCHLIKOVA, M. & IVANYI, P. (1974) DNA synthesis in the lymph node cells of mice. Differences associated with the H-2 system. Folia Biol. (Praha) 20, 68-71.

SALOMON, D.S., GIFT, V.D. & PRATT, R.M. (1979) Corticosterone levels during midgestation in the maternal plasma and fetus ot cleft palate sensitive and resistant mice. Endocrinology 104, 154.

SALOMON, D.S. & PRATT, R.M. (1976) Glucocorticoid receptors in murine embryonic facial mesenchyme cells. Nature 264, 174-177.

SALOMON, D.S. & PRATT, R.M. (1978) Inhibition of growth in vitro by gluco-corticoids in mouse embryonic facial mesenchyme cells. J. Cell Physiol. 97, 315-327.

SALOMON, D.S. & PRATT, R.M. (1979) Involvement of glucocorticoids in the development of the secondary palate. Differentiation 13, 141-154.

SALOMON, D.S., ZUBAIRI, Y. & THOMPSON, E.B. (1978) Ontogeny and biochemical properties of glucocorticoid receptors in midgestation mouse embryos. J. Steroid Biochem. 9, 95-107.

SAXEN, I. (1973) Effects of hydrocortisone on the development in vitro of the secondary palate in two inbred strains of mice. Arch. Oral Biol. 18, 1469-1479.

SCATCHARD, G. (1949) The attraction of proteins for small molecules and ions. Ann. N.Y. Acad. Sci. 51, 660-672.

SHAH, R.M. & KILISTOFF, A. (1976) Cleft palate induction in hamster fetuses by glucocorticoid hormones and their synthetic analogues. J. Embryo. exp. Morph. 36, 101-108.

SHAPIRA, Y. & SHOSHAN, S. (1972) The effect of cortisone on collagen synthesis in the secondary palate in mice. Arch. Oral Biol. 17, 1699-1703.

SHIRE, J.G.M. (1974) Endocrine genetics of the adrenal gland. J. Endo-crinology 62, 173-207.

SMILEY, G.R. & KOCH, W.E. (1972) An in vitro and in vivo study of single palatal processes. Anat. Rec. 173, 405-416.

SNELL, G.D. (1968) The H-2 locus of the mouse: Observations and speculations concerning its comparative genetics and its polymorphism. Folia Biol. (Praha) 14, 335-358.

SNELL, G.D., CHERRY, M. & DEMAND, P. (1973) H-2: Its structure and similarity to HL-A. Transplant. Rev. 15, 3-25.

THRASH, C.R., HO, T-S. & CUNNINGHAM, D.D. (1974) Structural features of steroids which initiate proliferation of density-inhibited 3T3 mouse fibro-blasts. J. Biol. Chem. 249, 6099-6103.

TUAN-N-T, REKDAL, D. & BURTON, A.F. (1971) The uptake and metabolism of ^{3}H-corticosterone and flurometrically determined corticosterone in fetuses of several mouse strains. Biol. Neonate 18, 78-84.

TYAN, M.L. & MILLER, K.K. (1978) Genetic and environmental factors in cortisone induced cleft palate. Proc. Soc. Exp. Biol. Med. 158, 618-621.

TYLER, M.S. & KOCH, W.E. (1975) In vitro development of palatal tissues from embryonic mice. I. Differentiation of the secondary palate from day 12 mouse embryos. Anat. Rec. 182, 297-304.

TYLER, M.S. & KOCH, W.E. (1977) In vitro development of palatal tissues from embryonic mice (II and III). J. Embryo. exp. Morph. 38, 19-48.

TYLER, M.S. & PRATT, R.M. Exp. Embryo. & Morph., in press.

VENETIANER, A., BAJNOCZKY, K., GAL, A. & THOMPSON, E.B. (1978) Isolation and characterization of L cell variants with altered sensitivity to glucocorticoids. Som. Cell Genetics 4, 513-530.

WALKER, B.E. (1967) Induction of cleft palate in rabbits by several gluco-corticoids. Proc. Soc. exp. Biol. Med. 125, 1281-1284.

WALKER, B.E. & FRASER, F.R. (1957) The embryology of cortisone-induced cleft palate. J. Embryol. exp. Morph. 5, 201-209.

WILK, A.L., KING, C.T.G. & PRATT, R.M. (1978) Chlorcyclizine induction of cleft palate in the rat: Degradation of palatal glycosaminoglycans. Teratology 18, 199-209.

YAMAMOTO, M.R., GEHEING, U., STAMPFER, M.R. & SIBLEY, C.H. (1976) Genetic approaches to steroid hormone action. In: R.O. Greep (Ed.) Recent Progress In Hormone Research, Vol. 32, Academic Press, New York, pp. 3-52.

ZARROW, M.X., PHILPOTT, J.E. & DENENBERG, V.H. (1970) Passage of ^{14}C-4-corticosterone from the rat mother to the foetus and neonate. Nature 226, 1058-1059.

ZIMMERMANN, E.F., ANDREW, F. & KALTER, H. (1970) Glucocorticoid inhibition of RNA synthesis responsible for cleft palate in mice: A model. Proc. Natl. Acad. Sci. USA 67, 779-785.

ZIMMERMANN, E.F. & BOWEN, D. (1972a) Distribution and metabolism of triamcinolone acetonide in mice sensitive to its teratogenic effects. Teratology 5, 57-70.

ZIMMERMANN, E.F. & BOWEN, D. (1972b) Distribution and metabolism of triamcinolone acetonide in inbred mice with different cleft palate sensitivities. Teratology 5, 335-344.

FIBRONECTIN AND EARLY MAMMALIAN EMBRYOGENESIS

Jorma Wartiovaara and Antti Vaheri

Department of Electron Microscopy,
University of Helsinki,
Mannerheimintie 172, 00280 Helsinki 28, Finland

and

Department of Virology,
University of Helsinki,
Haartmaninkatu 3, 00290 Helsinki 29, Finland

1.   GENERAL PROPERTIES OF FIBRONECTIN

1.1   DIFFERENT FORMS OF FIBRONECTIN

The high molecular weight glycoprotein, fibronectin, is present in both soluble form in extracellular fluids and in insoluble form in interstitial connective tissue and in association with basement membranes (Table 1; for references, see Vaheri & Mosher, 1978; Yamada & Olden, 1978).

Soluble fibronectin, from the body fluids or the medium from adherent cell cultures, is a glycoprotein of 440,000 daltons and

TABLE 1:   THE DISTRIBUTION OF FIBRONECTIN IN VIVO

Location	References
Body fluids	
Plasma	1, 2, 3
Amniotic fluid	4
Allantoic fluid	5
Synovial fluid	6
Cerebrospinal fluid	7
Tissues	8, 9
Basement membranes	
Loose interstitial connective tissue matrix	
Periphery of individual cells	
Mesenchymal cells	
Smooth muscle cells	
Fat cells	
Proliferating fibroblasts	
Predominant locations	8, 9
Vascular walls	
Stroma of lymphatic tissue	
Submucosa	

References: 1, Mosesson & Unfleet, 1970; 2, Morrison et al, 1948; 3, Ruoslahti & Vaheri, 1975; 4, Chen et al, 1976; 5, unpublished observations; 6, Vartio, T., Vaheri, A., von Essen, R., Isomäki, H. & Stenman, S. (unpublished); 7, Kuusela et al, 1978; 8, Stenman & Vaheri, 1978; 9, Linder et al, 1978.

is composed of two similar but not identical disulfide-bonded subunits of 220,000 daltons (Mosher, 1975; Mosesson, 1977; Kurkinen et al, 1980a). The concentration in normal human plasma is about 300 μg/ml (Mosesson & Umfleet, 1970). In amniotic (Chen et al, 1976), cerebrospinal (Kuusela et al, 1978) and rheumatoid synovial fluids (Vartio, T., Vaheri, A., von Essen, R., Isomäki, H. & Stenman, S. unpublished) its concentration as a proportion of total protein is higher, which might indicate local synthesis.

Matrix fibronectin has a wide distribution. It is found in vivo in association with basement membranes, in loose connective tissue matrix and around various types of individual cells (Table 1). Immunohistological studies have localised fibronectin predominantly to vascular walls, the stroma of lymphatic tissue and in the submucosa (Stenman & Vaheri, 1978). Fibronectin appears to be particularly abundant in developing tissues, as discussed in detail below.

Several types of adherent cells produce large amounts of fibronectin in culture. These include fibroblasts (Ruoslahti & Vaheri, 1974), cells of probable astroglial origin (Vaheri et al, 1976), endothelial cells (Birdwell et al, 1978; Jaffe & Mosher, 1978; Macarak et al, 1978), smooth muscle cells (see Vaheri & Mosher, 1978), macrophages (Alitalo et al, 1980a) and non-ectodermal epithelial cells such as those from gut (Quaroni et al, 1978), liver and kidney (Chen et al, 1977b; Voss et al, 1979). Most of the fibronectin synthesized by cultured cells is secreted in soluble form into the medium. In addition, normal adherent cells, unlike malignantly transformed cells in general (Hynes, 1976; Vaheri & Mosher, 1978), deposit fibronectin into a peri-cellular substrate-attached matrix (Fig.1) (Hynes, 1976; Vaheri et al, 1978a; Vaheri & Mosher, 1978), which may also contain various other glycoproteins and glycosaminoglycans. For example, cultured human fibroblasts produce a substrate-attached, cell-free, pericellular matrix constituted mainly of fibronectin, procollagens of types I and III, heparin sulfate proteoglycan and hyaluronic acid (Figs.2 and 3; Hedman et al, 1979).

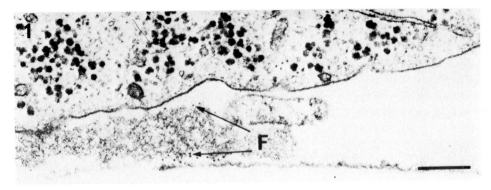

Figure 1: Normal human embryonic fibroblasts deposit fibronectin into a pericellular substratum-attached matrix as visualized here by immuno-ferritin
electron microscopy. F, ferritin markers. Bar 200 nm. (Hedman *et al*, 1978).

1.2 *INTERACTIONS OF FIBRONECTIN*

Fibronectin has been shown to have several interactions that
may be of biological relevance (Table 2). Purified soluble
fibronectin assembles spontaneously into filamentous structures
(Fig.4) (Vuento *et al*, 1980) that resemble the fibrillar fibronectin-containing structures seen in the pericellular matrix
material of, for example, cultured fibroblasts (Hedman *et al*,
1979). Soluble fibronectin also interacts with fibrin and heparin
in the cold (Mosesson & Umfleet, 1970; Ruoslahti & Vaheri, 1975;
Mosher, 1976; Stathakis & Mosesson, 1977) and also binds
effectively to denatured collagen (Engvall & Ruoslahti, 1977;
Dessau *et al*, 1978a; Engvall *et al*, 1978). These properties are
used evidently by fibronectin in the circulation, where it acts
as a nonspecific opsonin, promoting uptake of fibrin micro-
aggregates and collagen-containing microscopic particles by cells
of the reticuloendothelial system (Saba *et al*, 1978). Several
lines of evidence suggest that blood coagulation factor XIII_a
(plasma transglutaminase), a thrombin-activated cross-linking
enzyme, may mediate important interactions of fibronectin *in vivo*.
If blood is clotted at $+37^\circ C$, factor XIII_a cross-links fibronectin
to the α-chain of fibrin (Mosher, 1975) and in the presence of
reagents containing sulphydryl groups (Birckbichler & Patterson,
1978) fibronectin-fibronectin cross-links are also formed. Factor
XIII_a will also cross-link the pericellular matrix fibronectin of
cultured fibroblasts to form a very high molecular weight complex

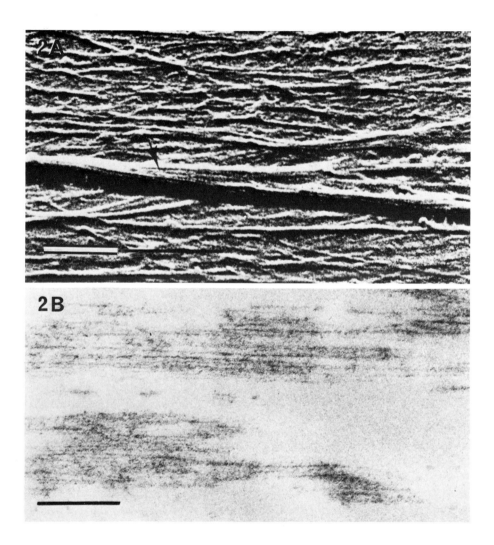

Figure 2: A. Scanning electron micrograph of pericellular matrix material attached to a substratum from which human embryonic fibroblasts were removed. Thicker fibers (arrow) seem to have a filamentous substructure. Bar 2 μm. B. Transmission electron micrograph of the same material as in Figure 2A. Bar 500 nm. (Hedman *et al*, 1979).

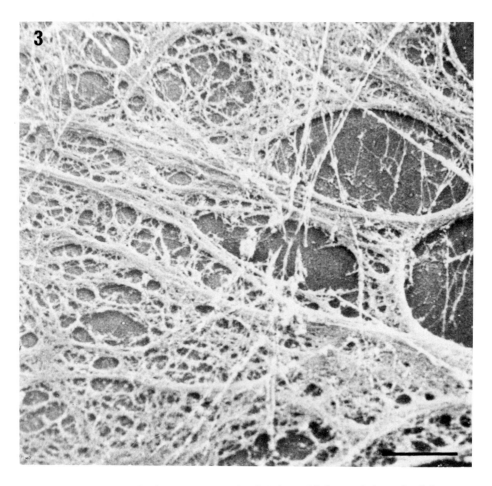

Figure 3: Scanning electron micrograph of intercellular matrix material produced by cultured human amniotic epithelial cells. Bar 200 nm. (K. Hedman, K. Alitalo & A. Vaheri, unpublished).

(Keskioja *et al*, 1976a). More recently it has been found that the $\alpha 1$ chain of type I collagen can be cross-linked to fibronectin by factor XIII_a (Mosher *et al*, 1979). A close association may exist between fibronectin and collagen *in vivo* since purified collagenase solubilizes fibronectin from tissues (Bray, 1978). In addition to its attachment in the extracellular matrix, a connection of matrix fibronectin with intracellular cytoskeletal elements has also been suggested (see Yamada & Olden, 1978). For

Figure 4: Purified soluble human fibronectin assembled spontaneously into filamentous structures. Prepared as in Vuento *et al*. (1980). Bar 240 nm.

example, fibronectin binds to columns of actin-agarose (Keski-Oja *et al*, 1979), and colinear distribution of intracellular actin filaments with extracellular fibronectin strands has been detected by immunofluorescence studies (Hynes & Destree, 1978; Heggeness *et al*, 1978). Immunoelectronmicroscopy has revealed that structures containing the actin and fibronectin molecules are separated by less than 8-22 nm across the cell membrane (Singer, 1979). A transmembrane protein has been postulated therefore to link the extracellular fibronectin fibrils with the intracellular actin microfilaments (Lloyd, 1979). Disruption of the microfilament system by cytochalasin treatment leads to loss of surface-associated fibronectin (Ali & Hynes, 1977; Kurkinen *et al*, 1978; Nielsen & Puck, 1980) and suggests that a close association between microfilaments and fibronectin has a role to play in the maintenance of cell shape. A role for fibronectin

TABLE 2: INTERACTIONS OF FIBRONECTIN

Interaction	Comments	References
Binding to		
Fibronectin	Assembly into filamentous polymers	1
Collagen	Defined binding sites in collagen; strong binding to gelatin	2, 3, 4
Fibrin (in the cold)	Bound to fibrin clot	5, 6, 7
Heparin (in the cold)		8
Cell surfaces		9, 16
Certain bacteria	Agglutinates S. aureus	10
Actin	Binds to actin-agarose	11
DNA	Binds to DNA-agarose	12, 13
Susceptible to		
Disulfide bonding	Soluble fibronectin is a dimer; in matrix is more extensively disulfide-bonded	14, 15
Proteolytic enzymes		9
Transglutaminase (factor XIII_a)	Both soluble and matrix forms cross-linked	17, 18, 19

References: 1, Vuento *et al*, 1980; 2, Dessau *et al*, 1978a; 3, Engvall & Ruoslahti, 1977; 4, Engvall *et al*, 1978; 5, Mosesson & Umfleet, 1970; 6, Mosher, 1976; 7, Ruoslahti & Vaheri, 1975; 8, Stathakis & Mosesson, 1977; 9, see: Vaheri & Mosher, 1978; Mosher, 1980; Yamada & Olden, 1978; 10, Kuusela, 1978; 11, Keski-Oja *et al*, 1979; 12, Zardi *et al*, 1979; 13, Parsons *et al*, 1979; 14, Keski-Oja *et al*, 1977; 15, Hynes & Destree, 1977; 16, see: Hynes 1976; Vaheri & Mosher, 1978; 17, Keski-Oja *et al*, 1976a; 18, Mosher, 1976; 19, Mosher *et al*, 1979.

in changes in cell motility and migratory behaviour has been described in cell cultures of normal and transformed fibroblastic cells (Ali & Hynes, 1978), differentiating neural crest cells (Sieber-Blum *et al*, 1978) and chick corneal cells (Kurkinen *et al*, 1979) *in vivo*.

Fibronectin, either in soluble or matrix form, is readily degraded by a variety of proteolytic enzymes (for references, see

Hynes, 1976; Vaheri & Mosher, 1978). When fibronectin is compared with other plasma and matrix proteins, it appears as a
relatively selected substrate of the neutral tissue proteases,
chymase and cathepsin G (Vartio, T., Seppä, H. & Vaheri, A.,
unpublished). After limited digestion of fibronectin with
plasmin or trypsin, the disulfide-bonded dimer of two 220,000
dalton subunits is cleaved into fragments of 200,000 daltons.
These fragments are no longer disulfide-bonded and, in the case
of pericellular matrix fibronectin, are released into the medium
(Keski-Oja et al, 1976b; Jilek & Hörmann, 1977; Ruoslahti et al,
1979; Wagner & Hynes, 1979). Thus, the intersubunit disulfide
bridges may be involved critically in the interactions of fibronectin with cell surfaces.

1.3   FIBRONECTIN AND MALIGNANT CELLS

Malignantly transformed cells, unlike many types of normal
cell in culture, fail generally to deposit a fibronectin-
containing, pericellular matrix (Table 3). The reason for the

TABLE 3:   ALTERATIONS IN PERICELLULAR
FIBRONECTIN IN CELL CULTURES

Event or treatment	Observation	Reference
Malignant transformation	Loss or decrease of fibronectin	1, 2, 3
Mitosis	As above	4, 5
Plasmin or activated plasminogen	As above	6
Cytochalasin B	As above	7, 8
Epidermal growth factor	Increase in fibronectin of mouse 3T3 cells	9
Glucocorticoid hormones	Increase in fibronectin both in normal and trans- formed fibroblast cultures	10, 11

References: 1, Hynes, 1976; 2, Vaheri & Mosher, 1978; 3, Yamada
& Olden, 1978; 4, Hynes & Bye, 1974; 5, Stenman et al, 1977;
6, Blumberg & Robbins, 1975; 7, Ali & Hynes, 1977; 8, Kurkinen
et al, 1978; 9, Chen et al, 1977a; 10, Furcht et al, 1979b;
11, our unpublished observations.

defective association of fibronectin to transformed cell layers
is not understood (Vaheri & Mosher, 1978), but the defect may play
a significant role in transformation since addition of solubilized,
matrix fibronectin to transformed cells restores normal morphology, increased adhesion to the substratum (Yamada et al, 1976)
and organized microfilament bundles (Ali et al, 1977; Willingham
et al, 1977). Similarly, human sarcoma cells, when seeded on a
cell-free pericellular matrix isolated from fibroblast cultures,
flatten out and assume a morphology indistinguishable from that
of normal fibroblasts (Vaheri et al, 1978b).

2. FIBRONECTIN AND EARLY EMBRYOS

2.1 EARLY MOUSE DEVELOPMENT

Most of our knowledge of fibronectin in early mammalian
embryos is based on work done on mouse embryos and relevant
morphological events in the development of the mouse embryo will
be outlined briefly therefore. At the 8-cell stage, the
previously loose spherical cells of the embryo acquire a non-
spherical shape as the embryo becomes compacted. With the
ensuing 8- to 16-cell transition, some cells become localized
to the interior of the embryo to form the putative lineage of the
inner cell mass (ICM) (Barlow et al, 1972; Graham & Lehtonen,
1979). At the blastocyst stage the ICM develops into primary
ectoderm and endoderm and an embryonal basement membrane is
formed between these two germ layers. Extracellular material is
found also on the inner aspect of the trophectoderm in the preimplantation blastocyst (Schlafke & Enders, 1963; Nadijcka &
Hillman, 1974), and later the Reichert's membrane, a thick layer
of extracellular material, is laid down at this same site between
the parietal endoderm migrating over the outer trophoblast layer.
Other extraembryonic basement membranes develop later in the
chorion, in the amnion, and in the visceral yolk sac (cf. Snell
& Stevens, 1966). The third germ layer, the mesoderm, arises
from the primary embryonic ectoderm in the primitive streak area
of the 6½ day embryo. Early mesodermal derivatives include head
and heart mesenchyme, the somites and the allantois.

2.2   FIBRONECTIN IN THE EARLY MOUSE EMBRYO

The fibronectin in early mouse embryos has been studied by
the use of antibodies for localisation of the protein to specific
sites.   No fibronectin has been detected in the zygote, cleaving
or compacted embryo, the early blastocyst or the surrounding zona
pellucida (Fig.5).   First detection of fibronectin   occurs in
sections of implanting 4.5 day mouse embryos (Wartiovaara *et al*,
1978b) coincident with the formation of the primitive endoderm
around the inner cell mass.   The fibronectin localizes mainly
between the primitive endoderm and the ectoderm core (Fig.6) at
the site of the developing layer of extracellular material.
Similarly, Zetter and Martin (1978) detected fibronectin in
immunosurgically isolated ICM's in 4-day blastocysts maintained
in culture.   In later stages of endoderm development, when
primitive endoderm differentiates into the visceral endoderm
surrounding the egg cylinder and into the parietal endoderm
lining the inner aspect of the primitive yolk sac, fibronectin
is readily detectable in the extracellular material that forms a
layer under both endoderm types (Fig.7).   Especially strong fibronectin fluorescence is found at the end of the first week of
development in the Reichert's membrane between the parietal endoderm and the trophectoderm layers.

With the formation of the third germ layer, the mesoderm, in
the primitive streak area a separating layer of fibronectin
appears between the mesoderm and the ectoderm (Fig.8) from which
the former arises.   The mesoderm cells, unlike the ectoderm cells,
stain positive for fibronectin.   As the mesoderm expands around
the embryo, it is separated from both the adjacent ectodermal and
endodermal layers by fibronectin-containing extracellular
material.

With development of the extraembryonic membranes, fibronectin
becomes detectable between the cell layers forming the membranes.
In addition to the strong fibronectin staining in the Reichert's
membrane mentioned above, the allantois, the amnion and the
chorion also contain this protein (Fig.9).   In culture conditions, human amniotic epithelial cells secrete and deposit
fibronectin-containing extracellular material on their growth
substratum (Fig.3).   In the human post-partum amniotic membranes

244

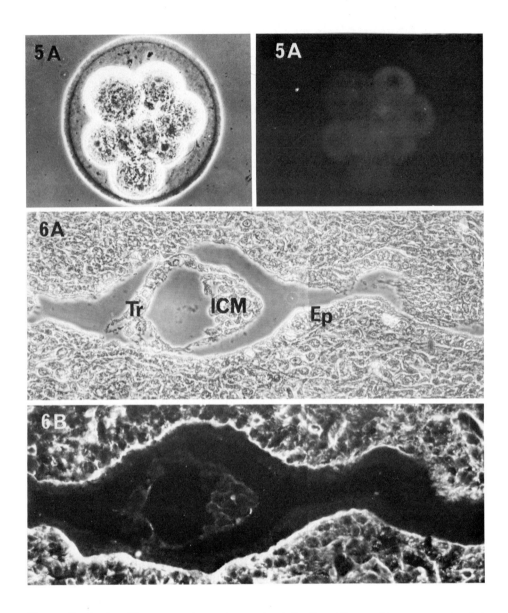

Figure 5: An 8-cell mouse embryo does not stain for fibronectin as seen in (A) phase-contrast and (B) immunofluorescence microscopy x 600. (Wartiovaara *et al*, 1979).

Figure 6: (A) Implanting 4½ day embryo attached to the uterine epithelium (Ep). Inner cell mass, ICM; trophectoderm, Tr. (B) Faint fibronectin fluorescence is seen as strands in the ICM, mainly on cells adjacent to the blastocoel cavity, and also on the blastocoel side of the trophectoderm. x 600. (Wartiovaara *et al*, 1979).

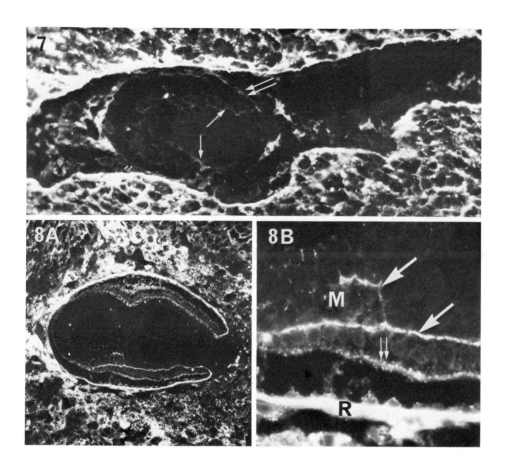

Figure 7: Implanted 5.5 day embryo in which fibronectin fluorescence forms a layer between the visceral endoderm and the ectoderm (single arrows) and a layer adjacent to the inner side of the trophectoderm (double arrow). x 600. (Wartiovaara *et al*, 1979).

Figure 8: Seven day embryo (A, x 700) and enlargement (B, x 2800) in which fibronectin fluorescence forms a band (single arrows) between endoderm and ectoderm and between mesoderm (M) and ectoderm. The Reichert's membrane stains strongly. The apical region of the visceral endoderm cells have granular fluorescence (double arrow). The trophoblastic giant cells around the embryo (A) show a heavy cytoplasmic staining pattern. (Wartiovaara *et al*, 1979).

fibronectin is detected between and under the amniotic epithelial cells and in the underlying acellular "compact layer" of the amniotic basement membrane (Alitalo *et al*, 1980b). These findings suggest that certain epithelial cells may deposit fibronectin into underlying basement membranes.

2.3 *IMPLANTATION AND FIBRONECTIN*

At implantation the embryo penetrates through the uterine epithelium and the underlying basement membrane into the loose uterine stroma. During this stage the trophoblast cells forming the outermost layer of the blastocyst secrete plasminogen activator (Sherman *et al*, 1976; Strickland *et al*, 1976 and this volume), a proteolytic enzyme thought to be involved in the invasion process. The breakdown of the uterine epithelial basement membrane appears to involve degradation of its fibronectin, presumably by activated plasmin. At the site of implantation, a decidual swelling reaction takes place in the uterine stroma. Simultaneously, a distinct change in its fibronectin pattern is seen. The network of fibronectin-containing extracellular material disappears and is replaced by the invading trophoblastic giant cells that show a very heavy granular cytoplasmic fluorescence (Figs.8,9).

3. FIBRONECTIN AND TERATOCARCINOMA CULTURES

3.1 *TERATOCARCINOMA AS MODEL OF EARLY DEVELOPMENT*

Teratocarcinomas are malignant tumours of embryonic or germ cell origin which contain embryonal carcinoma stem cells and usually a number of differentiated non-malignant derivatives of the stem cells (cf. Stevens, 1967; Damjanov & Solter, 1974). Because differentiation of embryonal carcinoma cells appears to parallel early mammalian embryogenesis in several aspects, mouse teratocarcinomas have been used extensively as a model for the study of developmental phenomena (Martin, 1975; Graham, 1977; Jacob, 1977). Human teratocarcinoma lines are also available

Figure 9: In 7½ day embryo, prominent fibronectin fluorescence is seen in the Reichert's (R) membrane, the amnion (A) and the chorion (C) as well as in the trophoblastic giant cells (GC). A, Nomarski optics, x 450. B, Indirect immunofluorescence, x 450. (Wartiovaara *et al*, 1979).

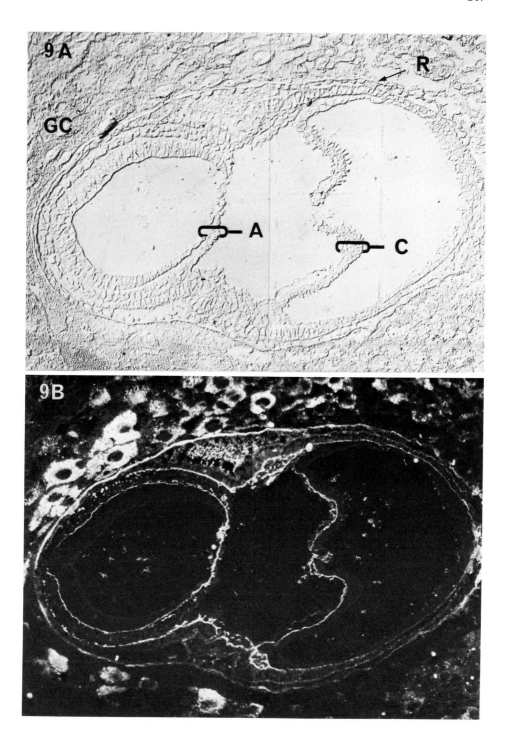

(Hogan *et al*, 1977; Holden *et al*, 1977; Zeuthen *et al*, 1980).
Embryonal carcinoma cells cultured under non-adherent conditions
can form aggregates called embryoid bodies that are analogous in
their structure to the inner cell mass of the 5 day mouse
embryo (Fig.10). Some clonal lines termed nullipotent embryonal
carcinomas lack the capacity to differentiate. Multipotent cell
lines can differentiate both in embryoid bodies and in monolayer
culture in an endodermal direction, mimicking the formation of
endoderm in the blastocyst. In this process, extracellular
material is deposited under the endodermal cells (Martin, 1975).

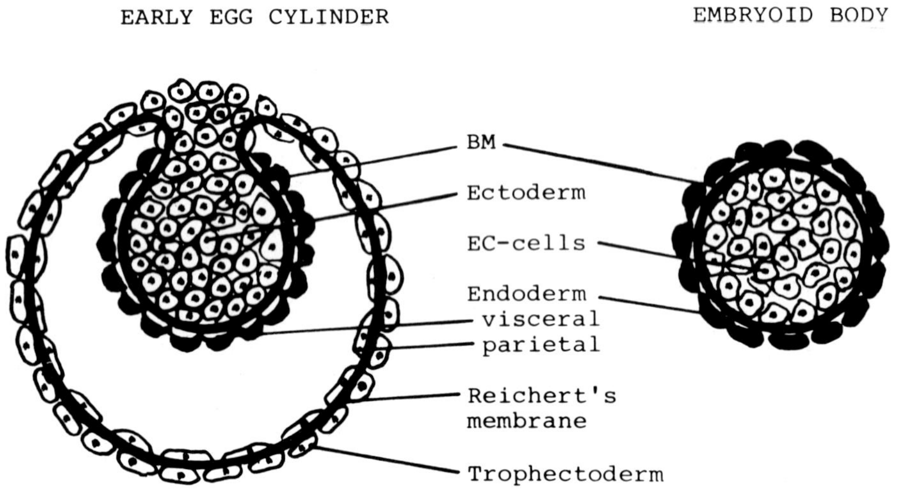

EARLY EGG CYLINDER   EMBRYOID BODY

BM
Ectoderm
EC-cells
Endoderm
visceral
parietal
Reichert's
membrane
Trophectoderm

Figure 10: Analogous structures in the early egg cylinder of the mouse embryo
and in the teratocarcinoma embryoid body. BM, basement membrane material;
EC-cells, embryonal carcinoma cells (After Wartiovaara *et al*, 1980).

3.2 EXPRESSION OF FIBRONECTIN BY TERATOCARCINOMA CELLS

Several lines of teratocarcinoma cells, both nullipotent and multipotent, have been tested for their capacity to produce and deposit fibronectin (Table 4). The results indicate that embryonal carcinoma cells of some teratocarcinoma lines including the nullipotent line SCC-1 and the multipotent line PSA 1 deposit fibronectin as judged by immunofluorescence (Zetter & Martin, 1978). However, stem cells of several other teratocarcinoma lines

TABLE 4:   FIBRONECTIN EXPRESSION IN TERATOCARCINOMA
CELLS

Cell line	Type	Cells	Fibronectin synthesis	deposition	References
SCC-1	Nullipotent	EC	+	- ; +	1, 2
PCC4/A/1	Multipotent	EC		-	
A6	Multipotent	EC	+	-	1
F9	Multipotent (?)	EC		-	2, 3
	Differentiated	End, EBA, Ml		+	3
PSA1	Multipotent	EC		+	2
	Differentiated	End, EBA		+	2
OC1531	Multipotent	EC		-	4, 5
	Differentiated	End, EBA		+	4, 5
	Differentiated	End, Ml		+	4
*PA 1	Multipotent (?)	EC		-	6
	Differentiated	End		+	6
PSA5-E	Differentiated	VE	+	+	1
PYS-1	Differentiated	PE	-	-	1
PYS-2	Differentiated	PE	-	+ ; -	2, 3, 7

References: 1, Wolfe et al, 1979; 2, Zetter & Martin, 1978; 3, Wariovaara et al, 1980; 4, Wartiovaara et al, 1978a; 5, Wartiovaara et al, 1978b; 6, Zeuthen et al, 1980; 7, Leivo et al (unpublished).

Abbreviations: EC: embryonal carcinoma; End: endodermal-type; EBA: embryoid body-like aggregate; Ml: monolayer; VE: visceral endoderm; PE: parietal endoderm.
* - human line.

including the OC1551 (Wartiovaara *et al*, 1978a), F9 (Zetter & Martin, 1978; Wartiovaara *et al*, 1980), PCC4/A/1 (Zetter & Martin, 1978) and A6 lines (Wolfe *et al*, 1979) do not stain for surface fibronectin. When the embryonal carcinoma cells differentiate into endoderm cells either in aggregates or in monolayer cultures, fibronectin is invariably deposited in pericellular matrix form (Table 4). In these embryoid body-like structures, the matrix becomes localized between the surface layer of endoderm cells and the core of embryonal carcinoma cells (Fig.11). In monolayer cultures the fibronectin-containing, extracellular material is found mostly as strands around the periphery of the endoderm-like cells (Fig.12).

Some of the stable teratocarcinoma-derived endoderm lines have also been studied with respect to fibronectin expression. PSA5-E, a visceral endoderm line, both produces and deposits fibronectin in extracellular form (Wolfe *et al*, 1979). Results on PYS, a parietal endoderm line, have been contradictory. This may in part be due to uncertainty about the exact equivalence of the PYS lines used, but another explanation should be considered. If PYS-2 cells are grown in culture media containing regular calf serum, pericellular fibronectin becomes deposited along the cell borders (Zetter & Martin, 1978; Wartiovaara *et al*, 1980). If, however, the cells are grown in fibronectin-free calf serum, no staining of matrix fibronectin is seen (Leivo, I., Alitalo, K., Vaheri, A., Risteli, L., Timpl, R. & Wartiovaara, J., unpublished results), a result similar to that obtained with cultured fibro-blasts in one report (Hayman & Ruoslahti, 1979). Metabolic labeling and immunoprecipitation with specific antisera also fail to demonstrate synthesis of fibronectin by the PYS-2 cells. Similar results have been obtained with the PYS-1 line (Wolfe *et al*, 1979). In the light of these results, previous data on matrix deposition of fibronectin by different cell lines (as detected by localizing methods) should be confirmed by metabolic studies. Another potential source of contradictory results can lie in the anti-fibronectin antibodies used. Unless the antisera have been produced with plasma fibronectin or have been affinity-purified, it is possible that activity against another matrix glycoprotein, termed laminin (Timpl *et al*, 1979), could lead to erroneous conclusions. Laminin has nearly the same electrophoretic mobility as monomeric fibronectin.

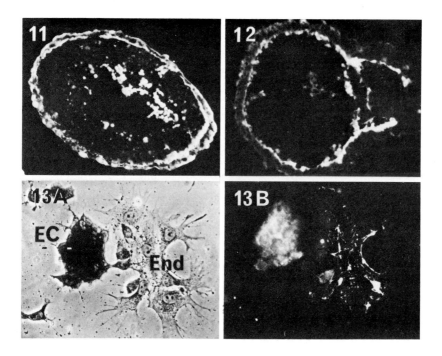

Figures 11 and 12: Ten-day, cystic embryoid body-like aggregate of mouse teratocarcinoma line F9 (11) and fifteen-day cystic embryoid body-like aggregate of line OC15S1 (12). Fibronectin fluorescence is localized to the surface layer of endoderm cells and the matrix between the endoderm and the core of embryonal carcinoma cells, which in cystic structures degenerates and stains non-specifically. x 650. (Wartiovaara *et al*, 1980).
Figure 13: F9 teratocarcinoma cells cultured for three days as a sparse monolayer in 3% fetal calf serum to allow endodermal differentiation. Formaldehyde fixation. x 250. (A) Staining for alkaline phosphatase gives strong reaction in embryonal carcinoma (EC) but not in endoderm-like (End) cells. (B) In indirect immunofluorescence for fibronectin, strong strand-like staining is seen especially in the periphery of the endodermlike cells while EC cells stain diffusely. x 1250. (Wartiovaara *et al*, 1980).

4. ROLES OF FIBRONECTIN

4.1 *FORMATION AND BREAKDOWN OF FIBRONECTIN-CONTAINING MATRIX*

Unlike the other proteins that have been identified in extracellular matrix, fibronectin is found both in the interstitial matrix and associated with basement membranes (Fig.14). Furthermore, soluble fibronectin can be deposited into a matrix form. This is known to occur in the fibrin-containing clot during the final stages of coagulation, generating a primary matrix (see Mosher, 1980), which presumably acts as a scaffold for the ingrowing connective tissue cells during the early stage of tissue repair.

As in the situation *in vivo*, so cells *in vitro* which synthesize interstitial types (I-III) of collagen (such as fibroblasts, smooth muscle cells, myoblasts and chondroblasts) secrete and deposit fibronectin (for references, see Vaheri & Mosher, 1978; Mosher, 1980). Studies with human embryonic skin fibroblasts seem to indicate that the deposition of fibronectin into the pericellular matrix may occur independently of that of collagenous proteins. Withdrawal of ascorbate or chelation of the ferrous ions, treatments that interfere with the processing and deposition of collagen, do not appreciably alter the deposition of fibronectin (unpublished observation). Purified bacterial collagenase may be used to remove collagenous proteins from the pericellular matrix around cultured fibroblasts without affecting the matrix fibronectin. Kinetic studies also suggest that the deposition of fibronectin occurs before that of procollagens (Vaheri *et al*, 1978a; Hedman *et al*, 1979; Kurkinen *et al*, 1980b). As mentioned above, purified soluble fibronectin tends to aggregate into filamentous structures, very similar to those observed by transmission electron microscopy in the pericellular fibronectin-containing matrix of cultured fibroblasts. Thus the deposition of fibronectin into pericellular matrix form may occur through a self-assembly process. Clearly, however, since only a small proportion of the fibronectin secreted by cultured fibroblasts is deposited, co-factors such as polyamines or sulfated

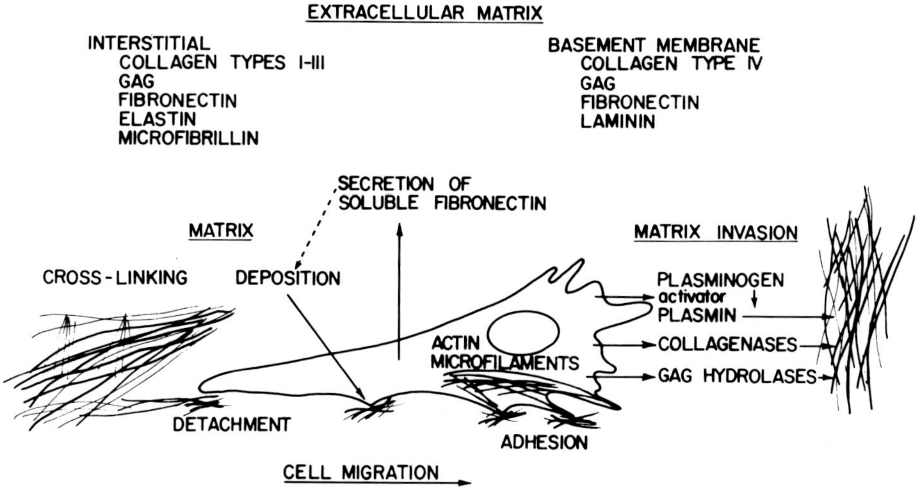

EXTRACELLULAR MATRIX

INTERSTITIAL
COLLAGEN TYPES I-III
GAG
FIBRONECTIN
ELASTIN
MICROFIBRILLIN

BASEMENT MEMBRANE
COLLAGEN TYPE IV
GAG
FIBRONECTIN
LAMININ

SECRETION OF SOLUBLE FIBRONECTIN

MATRIX

CROSS-LINKING DEPOSITION

MATRIX INVASION

PLASMINOGEN
activator
PLASMIN
COLLAGENASES
GAG HYDROLASES

ACTIN MICROFILAMENTS

DETACHMENT

ADHESION

CELL MIGRATION

Figure 14: Composition of extracellular matrix and involvement of fibronectin in cell-matrix interactions.

glycosaminoglycans or other cell surface molecules may be involved critically in fibronectin deposition.

Cell culture studies suggest that matrix fibronectin may be stabilized in part by disulfide bonding (Hynes & Destree, 1977; Keski-Oja et al, 1977). Further stabilization may occur through interaction of fibronectin with collagen, an interaction that has been shown to be stabilized under experimental conditions by sulfated glycosaminoglycans and by the cross-linking enzymes, transglutaminases.

Experiments, both with cell culture and *in vivo* indicate that fibronectin may play a key role in the organization of newly formed interstitial matrix. When human fibroblasts are left in culture for several weeks, the matrix changes (Furcht et al, 1979a; Mosher, 1980): the fibrils become thicker and according to immunoelectron microscopy contain both fibronectin and type I procollagen. Thus, two types of fibrils can be distinguished in culture. The first is thin (diameter 10-20 nm) and nonperiodic, contains fibronectin and types I and III procollagen. These thin fibrils are destroyed by trypsinization, and fibronectin apparently is their principal structural element. These fibrils are probably the *in vitro* counterpart of microfibrils seen in embryonic tissues (Low, 1968). The second type of fibril (diameter 40-50 nm) has the periodicity of a collagen fibril and therefore must be composed of an ordered array of collagen molecules. Fibronectin is associated with the periodic fibrils at 70 nm intervals, probably at the sites of potential cleavage by tissue collagenase where fibronectin is known to bind to collagen (on the cyanogen bromide cleavage fragment 7 of the α1(I) chain: Kleinman *et al*, 1978). On the other hand, in an experimental system in mouse (Kurkinen *et al*, 1980b), the matrix proteins are deposited in the following order: fibronectin, procollagen type III, and procollagen type I. This is followed by collagen fibril formation and finally loss of fibronectin.

At the level of light microscopy and immunofluorescence, fibronectin is detected in the basement membranes of most fully-formed organs (Stenman & Vaheri, 1978). However, according to recent immunoelectron microscopic studies (Oberley *et al*, 1979; Couchman *et al*, 1979), fibronectin is located between the lamina rara of the basement membrane and the abutting surfaces of cells rather than in the basement membrane itself. Conceivably, fibronectin may mediate the interaction between cell surface and basement membrane and may play a key role in anchorage and spreading of epithelial cells.

Phenomena such as cell migration and invasion, both during normal and pathological conditions, imply breakdown of fibronectin-containing matrices (Fig.14). How this is effected *in vivo* is not known, but proteolytic cleavage seems a likely possibility. Fibronectin is cleaved readily at specific protease-sensitive sites by various serine proteases such as trypsin and plasmin (see Vaheri & Mosher, 1978) and by larger doses of thrombin (Furie & Rifkin, 1980). The selective sensitivity of fibronectin to the neutral tissue proteases, chymase, elastase and cathepsin G, is of special interest because of their large quantities in tissues, in adult organs at least (Vartio *et al*,

unpublished). Nonproteolytic disintegration of the fibronectin-
containing matrix is also a real alternative as shown by model
studies. Fibronectin may be eluted from collagen by relatively
small concentrations of various compounds containing amino groups
(Vuento & Vaheri, 1978). The effects of hyaluronic acid are also
of interest since glycosaminoglycans have been associated with a
variety of developmental events (Cohn et al, 1977; Lash & Vasan,
1978; Ekblom et al, 1979). Although hyaluronic acid apparently
does not compete for binding of heparin to fibronectin
(Ruoslahti & Engvall, 1980; Yamada et al, 1980), it is an effective inhibitor of the precipitation of native collagen induced by
fibronectin and heparin (Jilek & Hörmann, 1979). Conceivably,
the destablizing effect of hyaluronic acid on heparin-fibronectin
complexes may be important for cell migration.

4.2 FIBRONECTIN IN EARLY CYTODIFFERENTIATION AND MORPHOGENESIS

The development of fibronectin in the early mouse embryo
seems to be associated closely in its appearance and localization
with that of extracellular matrix material, and with the cells
bordering it (Table 5). This is evident first in the formation
of the three germ layers, but also at later stages of development
and in the teratocarcinoma system.

The time of onset of fibronectin expression remains still an
open question in the light of the lack of metabolic studies.
Zetter and Martin (1978) detected fibronectin by immunofluorescence
on inner cell masses isolated immunosurgically from 4 day mouse
blastocysts kept in culture. In sectioned material, fibronectin
is first visible in 4.5 day embryos, mostly between the primitive
endoderm cells and the ectoderm core (Wartiovaara et al, 1978b).
As the ectoderm cells in sectioned embryos from later stages
stain negative for fibronectin (Wartiovaara et al, 1978b), Zetter
and Martin (1978) and Wolfe et al (1979) propose a transient
stage-specific role for fibronectin in ectoderm differentiation.
It still, however, remains possible that the early positive surface staining of the ICM cells seen by them in the isolated ICMs
could result from staining of superficial cells differentiating
into endoderm direction. The results from studies on teratocarcinomas described above (Table 4), speak in favour of a close

TABLE 5:   LOCALIZATION BY IMMUNOFLUORESCENCE OF
FIBRONECTIN IN EARLY MOUSE EMBRYO IN
CORRELATION WITH OTHER MATRIX GLYCOPROTEINS

Tissue or cell	Fibronectin	Type IV collagen	Laminin
Zygote	-	-	-
Uncompacted morula	-	-	-
Compacted morula	-	-	+
Blastocyst ICM	+	+	+
Ectoderm	-	-	+
Endoderm (VE, PE)	+	+	+
Mesoderm	+	+	+
Basement membranes, RM	+	+	+
Trophoblast	+	+	+

Abbreviations: ICM: inner cell mass;   VE: visceral endoderm;
PE: parietal endoderm; RM: Reichert's membrane

association between matrix fibronectin and the endodermal differentiation of multipotent embryonal cells. However, embryonal
carcinoma cells (equivalent to primary embryonic ectoderm?) of
some lines produce and even deposit fibronectin, although con-
flicting data exist on this point (Table 4). It is often difficult to rule out incipient differentiation of cultured multipotent embryonal carcinoma cells which can lead to false interpretation of synthesis data. Obviously this may be the case also
for the undifferentiated ICM cells of the early embryo.

Comparing the expression of fibronectin in early embryos
with that of other components of extracellular matrix (Table 5),
it is interesting to note that extracellular laminin is detectable
by immunofluorescence in the compacted morula before formation of
the inner cell mass (Leivo et al, 1980; Wartiovaara et al, 1980).
Laminin is subsequently found in the ICM cells and later in cells
of all three germ layers in addition to the basement membranes
between them. Type IV basement membrane collagen has been first
detected in the ICM of the 3 day early blastocyst both by
immunofluorescence (Leivo et al, 1980; Wartiovaara et al, 1980)
and immunoperoxidase (Adamson & Ayers, 1979) techniques. Later,
it has a fibronectin-type distribution in that it is not

detectable in the ectoderm but is present in the other two germ
layers as well as in all basement membrane structures. The
relationship of these three matrix components, fibronectin,
laminin and type IV collagen, is still quite unknown as well as
details on their biosynthesis and mechanism of deposition. The
teratocarcinoma system may well be suited for the study of these
questions, as already demonstrated in the case of collagen
synthesis (Adamson *et al*, 1979). We also lack knowledge on the
turnover and breakdown of extracellular matrices which must take
place continuously in the growing embryo. Mechanisms demonstrated
in vitro (Section 4.1) may apply also *in vivo*.

It is not possible to propose definitive functional roles for
fibronectin in embryogenesis although changes in fibronectin
expression are seen in a number of developmental situations
(Table 6). Some conjectures based on data from various systems
may be entertained, however. Fibronectin has been assigned a
structural role in building an extracellular scaffold at very
early stages in development. Aside from this structural role,
fibronectin may mediate adhesion between the cell surface and the
matrix. This mediation could occur via the ability of fibronectin
to bind both to matrix components, such as collagen, and cell
surface components, such as sulfated cell surface proteoglycans
of NIL cells (Perkins *et al*, 1979) or to distinct binding sites
on fibroblast surfaces (Gold *et al*, 1979; Hahn & Yamada, 1979a,
1979b; Ruoslahti & Hayman, 1979; Ruoslahti *et al*, 1979; Wagner &
Hynes, 1979). Fibronectin is known to increase the adhesion of
connective tissue cells to growth substratum and notably to
collagen (for reference: see Yamada & Olden, 1978; Grinnel, 1978;
Pena & Hughes, 1978). Fibronectin is associated also with the
surface of many epithelial cell lines (Chen *et al*, 1976; Quaroni
et al,1978; Alitalo *et al*, 1980b), it enhances adhesion and
spreading of hepatocytes (Höök *et al*, 1977) and of lens epithelial
cells on plastic substrata (Hughes *et al*, 1979) and might be
involved in cell-cell adhesion. Addition of fibronectin to transformed cells enhances their adhesion and restores a more normal
cell shape (Yamada *et al*, 1976; Ali *et al*, 1977). The possible
linkage of externally bound fibronectin to internal microfilaments
as described earlier, provides a route whereby fibronectin could

TABLE 6:   CHANGES IN FIBRONECTIN EXPRESSION
IN EMBRYOGENESIS

Event	Finding	References
Early embryo - mouse	Appearance between germ layers and in visceral and parietal endoderm and in mesoderm. Not produced by parietal endoderm in vitro	1, 2 3
Early embryo - chick	Appearance in ectoderm in germinal crescent	4
Later embryogenesis - chick	Disappearance from differentiating mesenchymal tissues, e.g. presumptive kidney, cartilage, muscle	5
Eye development - chick	Present in migrating corneal endothelial cells and in secondary corneal stroma	6
Limb formation - mouse and chick	Disappearance from presumptive cartilage concomitant with appearance of type II collagen	5, 7, 8
Kidney development - mouse	Disappearance from mesenchyme differentiating into epithelial tubules. Appearance in tubular basement membrane	9
Tooth development - mouse	Disappearance from mesenchyme differentiating into odontoblasts	10
Neural crest cell differentiation - quail	Appearance during formation of adrenergic cell aggregates	11

References: 1, Zetter & Martin, 1978; 2, Wartiovaara *et al*, 1979; 3, Jetten *et al*, 1979; 4, Critchley *et al*, 1979; 5, Linder *et al*, 1975; 6, Kurkinen *et al*, 1979; 7, Dessau *et al*, 1978b; 8, Silver & Pratt, 1979; 9, Wartiovaara *et al*, 1976; 10, Thesleff *et al*, 1979; 11, Sieber-Blum *et al*, 1978.

influence not just cell shape but also the orientation and occurrence of cell movement. Such actions could have great signi- ficance for cell differentiation and morphogenesis where adhesive cell contacts and cell locomotion appear to be of major importance.

CONCLUDING REMARKS

Fibronectin is unique as a matrix component as it is found both in interstitial connective tissue and in basement membranes and is also present in soluble form in tissue fluids. Fibro- nectin has several characteristic interactions including those with other matrix components and plasma proteins. The early appearance of fibronectin in germ layer formation and its almost ubiquituous presence in embryonic basement membranes are con- spicuous features. Results obtained with the teratocarcinoma model system have given analogous results. Considerable data have accumulated for cultured cells on the structure and meta- bolism of pericellular matrices containing fibronectin. The involvement of fibronectin in cell-substratum interactions, including cell adhesion and migration, as well as indirect *in vivo* evidence points to a role for fibronectin in cell differen- tiation and morphogenesis.

REFERENCES

ADAMSON, E.D. & AYERS, S.E. (1979) The localization and synthesis of some collagen types in developing mouse embryos. Cell 16, 953-965.

ADAMSON, E.D., GAUNT, S.J. & GRAHAM, C.F. (1979) The differentiation of teratocarcinoma stem cells is marked by the types of collagen which are synthesized. Cell 17, 469-476.

ALI, I.U. & HYNES, R.O. (1977) Effects of cytochalasin B and colchicine on attachment of a major surface protein of fibroblasts. Biochim. Biophys. Acta 471, 16-24.

ALI, I.U. & HYNES, R.O. (1978) Effects of LETS glycoprotein on cell motility. Cell 14, 439-446.

ALI, I.U., MAUTNER, V., LANZA, R. & HYNES, R.O. (1977) Restoration of normal morphology, adhesion and cytoskeleton in transformed cells by addition of a transformation-sensitive surface protein. Cell 11, 115-126.

ALITALO, K., HOVI, T. & VAHERI, A. (1980a) Fibronectin is produced by human macrophages. J. Exp. Med. 151, 602-618.

ALITALO, K., KURKINEN, M., VAHERI, A., KRIEG, T. & TIMPL, R. (1980b) Extra- cellular matrix components synthesized by human amniotic epithelial cells in culture. Cell 19, 1053-1062.

BARLOW, P.W., OWEN, D.A.S. & GRAHAM, C.F. (1972) DNA-synthesis in the pre-implantation mouse embryos. J. Embryol. Exp. Morph. 27, 431-445.

BIRCKBICHLER, P.J. & PATTERSON, M.K. Jr. (1978) Cellular transglutaminase, growth and transformation. Ann. N.Y. Acad. Sci. 312, 354-365.

BIRDWELL, C.R., GOSPODAROWICZ, D. & NICOLSON, G.L. (1978) Identification, localization, and role of fibronectin in cultured bovine endothelial cells. Proc. Nat. Acad. Sci. U.S.A. 75, 3273-3277.

BLUMBERG, P.M. & ROBBINS, P.W. (1975) Effect of proteases on activation of resting chick embryo fibroblasts and cell surface proteins. Cell 6, 137-147.

BRAY, B.A. (1978) Cold-insoluble globulin (fibronectin) in connective tissues of adult human lung and in trophoblast basement membrane. J. Clin. Invest. 62, 745-752.

CHEN, A.B., MOSESSON, M.W. & SOLISH, G.I. (1976) Identification of the cold-insoluble globulin of plasma in amniotic fluid. Am. J. Obstet. Gynecol. 125, 958-961.

CHEN, L.B., GUDOR, R.C., SUN, T., CHEN, A.B. & MOSESSON, M.W. (1977a) Control of a cell surface major glycoprotein by epidermal growth factor. Science 197, 776-778.

CHEN, L.B., MAITLAND, N., GALLIMORE, P.H. & McDOUGALL, J.K. (1977b) Detection of the large external transformation-sensitive protein on some epithelial cells. Exp. Cell. Res. 106, 39-46.

COHN, R.H., BANERJEE, S.D. & BERNFIELD, M.R. (1977) Basal lamina of embryonic salivary epithelia. Nature of glucosaminoglycan and organization of extra-cellular materials. J. Cell Biol. 73, 464-478.

COUCHMAN, J.R., GIBSON, W.T., THOM, D., WEAVER, A.C., REES, D.A. & PARISH, W.E. (1979) Fibronectin distribution in epithelial and associated tissues of the rat. Arch. Dermatol. 266, 295-310.

CRITCHLEY, D.R., ENGLAND, M.A., WAKELY, J. & HYNES, R.O. (1979) Distri-bution of fibronectin in the ectoderm of gastrulating chick embryos. Nature, London, 280, 498-500.

DAMJANOV, I. & SOLTER, D. (1974) Experimental teratoma. Curr. Top. Pathol. 59, 69-130.

DESSAU, W., ADELMANN, B.C., TIMPL, R. & MARTIN, G.R. (1978a) Identification of the sites in collagen α-chains that bind serum anti-gelatin factor (cold-insoluble globulin). Biochem. J. 169, 55-59.

DESSAU, W., SASSE, J., TIMPL, R., JILEK, F. & VON DER MARK, K. (1978b) Synthesis and extracellular deposition of fibronectin in chondrocyte cultures. Response to the removal of extracellular cartilage matrix. J. Cell Biol. 79, 342-355.

EKBLOM, P.P., LASH, J.W., LEHTONEN, E., NORDLING, S. & SAXEN, L. (1979) Inhibition of morphogenetic cell interactions by 6-diazo-5-oxo-norleucine (DON). Exp. Cell Res. 121, 121-126.

ENGVALL, E. & RUOSLAHTI, E. (1977) Binding of soluble form of fibroblast surface protein, fibronectin, to collagen. Int. J. Cancer 20, 1-5.

ENGVALL, E., RUOSLAHTI, E. & MILLER, E.J. (1978) Affinity to collagens of different genetic types and to fibrinogen. J. Exp. Med. 147, 1584-1595.

FURCT, L.T., MOSHER, D.F., WENDELSCHAFER-CRABB, G. & FOIDART, J.M. (1979a) In vitro evolution of an extracellular fibronectin and collagen matrix on human fibroblasts. J. Supramol. Struct. Suppl. 3, 197.

FURCHT, L.T., MOSHER, D.F., WENDELSCHAFER-CRABB, G., WOODBRIDGE, P.A. & FOIDART, J.M. (1979b) Dexamethone-induced accumulation of a fibrin and collagen extracellular matrix in transformed human cells. Nature, London, 277, 393-395.

FURIE, M.B. & RIFKIN, D.B. (1980) Proteolytically-derived fragments of human plasma fibronectin and their localization within the intact molecule. J. Biol. Chem. (in press).

GOLD, L.I., GARCIA-PARDO, A., FRANGIONE, B., FRANKLIN, E.C. & PEARLSTEIN, E. (1979) Subtilisin and cyanogen bromide cleavage products of fibronectin that retain gelatin-binding activity. Proc. Nat. Acad. Sci. U.S.A. 76, 4803-4807.

GRAHAM, C.F. (1977) Teratocarcinoma cells and normal embryogenesis. In: M.J. Sherman (Ed.), Concepts in Mammalian Embryogenesis, MIT Press, Cambridge, Mass., pp.315-394.

GRAHAM, C.F. & LEHTONEN, E. (1979) Formation and consequence of cell patterns in preimplantation mouse development. J. Embryol. Exp. Morph. 49, 277-294.

GRINNEL, F. (1978) Cellular adhesiveness and extracellular substrata. Int. Rev. Cytol. 55, 65-144.

HAHN, L-H.E. & YAMADA, K.M. (1979a) Identification and isolation of collagen-binding fragment of the adhesive glycoprotein fibronectin. Proc. Nat. Acad. Sci. U.S.A. 76, 1160-1163.

HAHN, L-H.E. & YAMADA, K.M. (1979b) Isolation and biological characterization of active fragments of the adhesive glycoprotein fibronectin. Cell 18, 1043-1051.

HAYMAN, E.G. & RUOSLAHTI, E. (1979) Distribution of fetal bovine serum fibronectin and endogenous rat cell fibronectin in extracellular matrix. J. Cell Biol. 83, 10-14.

HEDMAN, K., VAHERI, A. & WARTIOVAARA, J. (1978) External fibronectin of cultured human fibroblasts is predominantly a matrix protein. J. Cell Biol. 76, 748-760.

HEDMAN, K., KURKINEN, M., ALITALO, K., VAHERI, A., JOHANSSON, S. & HÖÖK, M. (1979) Isolation of the pericellular matrix of human fibroblast cultures. J. Cell. Biol. 81, 83-91.

HEGGENESS, M.H., ASH, J.F. & SINGER, S.J. (1978) Transmembrane linkage of fibronectin to intracellular actin-containing filaments in cultured human fibroblasts. Ann. N.Y. Acad. Sci. 312, 414-417.

HOGAN, B., FELLOUS, M., AVNER, P.R. & JACOB, F. (1977) Isolation of a human teratoma cell line which expressed Fg antigen. Nature, London, 270, 515-518.

HOLDEN, S., BERNARD, O., ARTZT, K., WHITMORE, W.F. & BENNETT, D. (1977) Human and mouse embryonal carcinoma cells in culture share and embryonic antigen (Fg). Nature, London, 270, 518-520.

HÖÖK, M., RUBIN, K., OLDBERG, Å., ÖBRINK, B. & VAHERI, A. (1977) Cold-insoluble globulin mediates the adhesion of rat liver cells to plastic petri dishes. Biochem. Biophys. Res. Commun. 79, 726-733.

HUGHES, R.C., MILLS, G., COURTOIS, Y. & TASSIN, J. (1979) Role of fibronectin in the adhesiveness of bovine lens epithelial cells. Biol. Cellulaire 36, 321-330.

HYNES, R.O. (1976) Cell surface proteins and malignant transformation. Biochim. Biophys. Acta 458, 73-107.

HYNES, R.O. & BYE, J.B. (1974) Density and cell cycle dependence of cell surface protein in hamster fibroblasts. Cell 3, 113-120.

HYNES, R.O. & DESTREE, A. (1977) Extensive disulfide bonding at the mammalian cell surface. Proc. Nat. Acad. Sci. U.S.A. 74, 2855-2859.

HYNES, R.O. & DESTREE, A.T. (1978) Relationships between fibronectin (LETS protein) and actin. Cell 15, 875-886.

JACOB, F. (1977) Mouse teratocarcinomas and embryonic antigens. Immunol. Rev. 33, 3-32.

JAFFE, E.A. & MOSHER, D.F. (1978) Synthesis of fibronectin by cultured human endothelial cell. J. Exp. Med. 147, 1779-1791.

JETTEN A.M., JEFFEN, M.E.R. & SHERMAN, M.J. (1979) Analysis of cell surface and secreted proteins of primary cultures of mouse extraembryonic membranes. Devel. Biol. 70, 89-104.

JILEK, F. & HÖRMANN, H. (1977) Cold-insoluble globulin: Plasminolysis of cold-insoluble globulin. Hoppe-Seyler's Z. Physiol. Chem. 358, 133-136.

JILEK, F. & HÖRMANN, H. (1979) Fibronectin (cold-insoluble globulin): Influence of heparin and hyaluronic acid on the binding of native collagen. Hoppe-Seyler's Z. Physiol. Chem. 360, 597-603.

KESKI-OJA, J., MOSHER, D.F. & VAHERI, A. (1976a) Cross-linking of a major fibroblast surface-associated glycoprotein (fibronectin catalyzed by blood coagulation factor XIII. Cell 9, 29-35.

KESKI-OJA, J., VAHERI, A. & RUOSLAHTI, E. (1976b) Fibroblast surface antigen (SF): The external glycoprotein lost in proteolytic stimulation and malignant transformation. Int. J. Cancer 17, 261-269.

KESKI-OJA, J., MOSHER, D.F. & VAHERI, A. (1977) Dimeric character of fibronectin, a major cell surface-associated glycoprotein. Biochem. Biophys. Res. Commun. 74, 699-706.

KESKI-OJA, J., SEN, A. & TODARO, G.J. (1979) Soluble forms of fibronectin bind to purified actin molecules in vitro. J. Cell Biol. 83, 48a.

KLEINMAN, H.K., McGOODWIN, E.B., MARTIN, G.R., KLEBE, R.J., FIETZEK, P.P. & WOOLLEY, D.E. (1978) Localization of the binding site for cell attachment in the α1(I) chain of collagen. J. Biol. Chem. 253, 5642-5646.

KURKINEN, M., WARTIOVAARA, J. & VAHERI, A. (1978) Cytochalasin B releases a major surface-associated glycoprotein, fibronectin, from cultured fibroblasts. Exp. Cell Res. 111, 127-137.

KURKINEN, M., ALITALO, K., VAHERI, A., STENMAN, S. & SAXEN, L. (1979) Fibronectin in the development of embryonic chick eye. Devel. Biol. 69, 589-600.

KURKINEN, M., VARTIO, T. & VAHERI, A. (1980a) Polypeptides of human fibronectin are similar but not identical. Biochim. Biophys. Acta (in press).

KURKINEN, M., VAHERI, A., ROBERTS, P.J. & STENMAN, S. (1980b) Sequential appearance of fibronectin and collagen in experimental granulation tissue. Lab. Invest (in press).

KUUSELA, P. (1978) Fibronectin binds to Staphylococcus aureus. Nature 276, 718-720.

KUUSELA, P., VAHERI, A., PALO, J. & RUOSLAHTI, E. (1978) Demonstration of fibronectin in human cerebrospinal fluid. J. Lab. Clin. Med. 92, 595-601.

LASH, J.W. & VASAN, N.S. (1978) Somite chondrogenesis in vitro: stimulation by exogenous extracellular matrix components. Devel. Biol. 66, 151-171.

LEIVO, I., VAHERI, A., TIMPL, R. & WARTIOVAARA, J. (1980) Appearance and distribution of collagens and laminin in the early mouse embryo. Devel. Biol. 476, 100-114

LINDER, E., VAHERI, A., RUOSLAHTI, E. & WARTIOVAARA, J. (1975) Distribution of fibroblast surface antigen in the developing chick embryo. J. Exp. Med. 142, 41-49.

LINDER, E., STENMAN, S., LEHTO, V.-P. & VAHERI, A. (1978) Distribution of fibronectin in human tissues and relationship to other connective tissue components. Ann. N.Y. Acad. Sci. 312, 151-159.

LLOYD, C. (1979) Fibronectin: A function at the junction. Nature, London, 279, 473-474.

LOW, F.N. (1968) Extracellular connective tissue fibrils in the chick embryo. Anat. Rec. 160, 93-108.

MACARAK, E.J., KIRBY, E., KIRK, T., & KEFALIDES, N.A. (1978) Synthesis of cold-insoluble globulin by cultured calf endothelial cells. Proc. Nat. Acad. Sci. U.S.A. 75, 2621-2625.

MARTIN, G.R. (1975) Teratocarcinomas as a model system for the study of embryogenesis and neoplasia. Cell 5, 229-243.

MORRISON, P.R., EDSALL, J.T. & MILLER, S.G. (1948) Preparation and properties of serum and plasma proteins. XVIII. The separation of purified fibrinogen from fraction I of human plasma. J. Amer. Chem. Soc. 70, 3103-3108.

MOSESSON, M.W. (1977) Cold-insoluble globulin (cig), a circulating cell surface protein. Thromb. Haemost. 38, 742-750.

MOSESSON, M.W. & UMFLEET, R.A. (1970) The cold-insoluble globulin of human plasma. J. Biol. Chem. 245, 5728-5736.

MOSHER, D.F. (1975) Cross-linking of cold-insoluble globulin by fibrin-stabilizing factor. J. Biol. Chem. 250, 6614-6621.

MOSHER, S.F. (1976) Action of fibrin-stabilizing factor on cold-insoluble globulin and α_2-macroglobulin in clotting plasma. J. Biol. Chem. 251, 1639-1645.

MOSHER, D.F. (1980) Fibronectin. Progr. Hemost. Thromb. (in press).

MOSHER, D.F., SCHAD, P.E. & KLEINMAN, H.K. (1979) Cross-linking of fibronectin to collagen by blood coagulation factor XIIIa. J. Clin. Invest. 64, 781-787.

NADIJCKA, M. & HILLMAN, N. (1974) Ultrastructural studies of the mouse blastocyst substages. J. Embryol. Exp. Morph. 32, 675-695.

NIELSON, S.E. & PUCK, T.T. (1980) Deposition of fibronectin in the course of reverse transformation of Chinese hamster ovary cells by cyclic AMP. Proc. Nat. Acad. Sci. U.S.A. 77, 985-989.

OBERLEY, T.D., MOSHER, D.F. & MILLS, M.D. (1979) Localization of fibronectin within the renal glomerulus and its production by cultured glomerular cells. Am. J. Path. 96, 651-663.

PARSONS, R.G., TODD, H.D. & KOWAL, R. (1979) Isolation and identification of a human serum fibronectin-like protein elevated during malignant disease. Cancer Res. 39, 4341-4345.

264

PENA, S.D.J. & HUGHES, R.C. (1978) Fibronectin-plasma membrane interactions in the adhesion and spreading of hamster fibroblasts. Nature, London, 276, 80-83.

PERKINS, M.E., TAE, H.Ji. & HYNES, R.O. (1979) Cross-linking of fibronectin to sulfated proteoglycans at the cell surface. Cell, 16, 941-952.

QUARONI, A., ISSELBACHER, K.J. & RUOSLAHTI, E. (1978) Fibronectin synthesis by epithelial crypt cells of rat small intestine. Proc. Nat. Acad. Sci. U.S.A. 75, 5548-5552.

RUOSLAHTI, E. & ENGVALL, E. (1980) Effect of glycosaminoglycans on complexing of fibronectin and collagen. Int. J. Cancer. (in press).

RUOSLAHTI, E. & HAYMAN, E.G. (1979) Two active sites with different characteristics in fibronectin. FEBS Lett. 97, 221-224.

RUOSLAHTI, E. & VAHERI, A. (1974) Novel human serum protein from fibroblast plasma membrane. Nature, London, 248, 789-791.

RUOSLAHTI, E. & VAHERI, A. (1975) Interaction of soluble fibroblast surface (SF) antigen with fibrinogen and fibrin. Identity with cold insoluble globulin of human plasma. J. Exp. Med. 141, 497-501.

RUOSLAHTI, E., HAYMAN, E.G., KUUSELA, P., SHIVELY, J.E. & ENGVALL, E. (1979) Isolation of a tryptic fragment containing the collagen-binding site of plasma fibronectin. J. Biol. Chem. 254, 6054-6059.

SABA, T.M., BLUMENSTOCK, F.A., SCOVILL, W.A. & BERNARD, H. (1978) Cryo-precipitate reversal of opsonic α_2-surface binding glycoprotein deficiency in septic surgical and trauma patients. Science 201, 622-624.

SCHLAFKE, S. & ENDERS, A. (1963) Observations on the fine structure of the rate blastocyst. J. Anat. 97, 353-360.

SHERMAN, M.I., STRICKLAND, S. & REICH, E. (1976) Differentiation of early mouse embryonic and teratocarcinoma cells in vitro: plasminogen activator production. Cancer Res. 36, 4208-4216.

SIEBER-BLUM, M., SIEBER, F., YAMADA, K.M. & COHEN, A.M. (1978) Cell surface fibronectin (FN) in cultures of differentiating neural crest cells. J. Cell Biol. 79, 31a.

SILVER, M.H. & PRATT, R.M. (1979) Distribution of fibronectin during limb and palate development in the mouse. J. Cell Biol. 83, 465a.

SINGER, I.I. (1979) The fibronexus: A transmembrane association of fibro-nectin-containing fibers and bundles of 5 nm micro-filaments in hamster and human fibroblasts. Cell 16, 675-685.

SNELL, G.D. & STEVENS, L.C. (1966) Early embryology. In: E.C. Green (Ed.), Biology of the Laboratory Mouse, McGraw-Hill, New York, pp.205-245.

STATHAKIS, N.E. & MOSESSON, M.W. (1977) Interactions among heparin, cold-insoluble globulin, and fibrinogen in formation of the heparin-precipitable fraction of plasma. J. Clin. Invest. 60, 855-865.

STENMAN, S. & VAHERI, A. (1978) Distribution of a major connective tissue protein, fibronectin, in normal human tissues. J. Exp. Med. 147, 1054-1064.

STENMAN, S., WARTIOVAARA, J. & VAHERI, A. (1977) Changes in the distribution of a major fibroblast protein, fibronectin, during mitosis and interphase. J. Cell Biol. 74, 453-467.

STEVENS, L.C. (1967) The biology of teratomas. Advan. Morphogenesis 6, 1-81.

STRICKLAND, S., REICH, E. & SHERMAN, M.I. (1976) Plasminogen activator in early embryogenesis: Enzyme production by trophoblast and parietal endoderm. Cell 9, 231-240.

THESLEFF, I., STENMAN, S., VAHERI, A. & TIMPL, R. (1979) Changes in the matrix proteins, fibronectin and collagen, during differentiation of mouse tooth germ. Devel. Biol. 70, 116-126.

TIMPL, R., RHODE, H., ROBEY, P.G., RENNARD, S.I., FOIDART, J.M. & MARTIN, G.R. (1979) Laminin - a glycoprotein from basement membranes. J. Biol. Chem. 254, 9933-9937.

VAHERI, A. & MOSHER, D.F. (1978) High molecular weight, cell surface-associated glycoprotein (fibronectin) lost in malignant transformation. Biochim. Biophys. Acta 516, 1-25.

VAHERI, A., RUOSLAHTI, E., WESTERMARK, B. & PONTEN, J. (1976) A common cell-type specific surface-antigen in cultured human glial cells and fibroblasts: Loss in malignant cells. J. Exp. Med. 143, 64-72.

VAHERI, A., KURKINEN, M., LEHTO, V.-P., LINDER, E. & TIMPL, R. (1978a) Co-distribution of pericellular matrix proteins in cultured fibroblasts and loss in transformation: Fibronectin and procollagen. Proc. Nat. Acad. Sci. U.S.A. 75, 4944-4948.

VAHERI, A., ALITALO, K., HEDMAN, K., KESKI-OJA, J., KURKINEN, M. & WARTIOVAARA, J. (1978b) Fibronectin and the pericellular matrix in normal and transformed adherent cells. Ann. N.Y. Acad. Sci. 312, 444-449.

VOSS, B., ALLAM, S., RAUTERBERG, J., ULLRICH, K., GIESELMANN, V. & von FIGURA, K. (1979) Primary cultures of rat hepatocytes synthesize fibro-nectin. Biochem. Biophys. Res. Commun. 90, 1348-1354.

VUENTO, M. & VAHERI, A. (1978) Dissociation of fibronectin from gelatin-agarose by amino compounds. Biochem. J. 175, 333-336.

VUENTO, M., VARTIO, T., SARASTE, M., von BONSDORFF, C.-H. & VAHERI, A. (1980) Spontaneous and polyamine-induced formation of filamentous polymers from soluble fibronectin. Eur. J. Biochem. 105, 33-42

WAGNER, D.D. & HYNES, R.O. (1979) Domain structure of fibronectin and its relation to function. J. Biol. Chem. 254, 6746-6754.

WARTIOVAARA, J., STENMAN, S. & VAHERI, A. (1976) Changes in expression of fibroblast surface antigen (SFA) during cytodifferentiation and heterokaryon formation. Differentiation 5, 85-89.

WARTIOVAARA, J., LEIVO, I., VIRTANEN, I., VAHERI, A. & GRAHAM, C.F. (1978a) Appearance of fibronectin during differentiation of mouse teratocarcinoma in vitro. Nature, London, 272, 355-356.

WARTIOVAARA, J., LEIVO, I., VIRTANEN, I., VAHERI, A. & GRAHAM, C.F. (1978b) Cell surface and extracellular matrix glycoprotein fibronectin: Expression in embryogenesis and in teratocarcinoma differentiation. Ann. N.Y. Acad. Sci. 312, 132-141.

WARTIOVAARA, J., LEIVO, I. & VAHERI, A. (1979) Expression of the cell surface-associated glycoprotein, fibronectin, in the early mouse embryo. Devel. Biol. 69, 247-257.

WARTIOVAARA, J., LEIVO, I. & VAHERI, A. (1980) Matrix glycoproteins in early mouse development and in differentiation of mouse teratocarcinoma cells. In: S. Subtelny (Ed.), The Cell Surface: Mediator of Developmental Processes, Academic Press, New York, (in press).

WILLINGHAM, M.C., YAMADA, K.M., YAMADA, S.S., POUYSSEGUR, J. & PASTAN, I. (1977) Microfilament bundles and cell-shape are related to adhesiveness to substratum and are dissociable from growth-control in cultured fibroblasts. Cell 10, 375-380.

WOLFE, J., MAUTNER, V., HOGAN, B. & TILLY, R. (1979) Synthesis and retention of fibronectin (LETS protein) by mouse teratocarcinoma cells. Exp. Cell Res. 118, 63-71.

YAMADA, K.M. & OLDEN, K. (1978) Fibronectins: adhesive glycoproteins of cell surface and blood. Nature, London, 275, 179-184.

YAMADA, K.M., YAMADA, S.S. & PASTAN, I. (1976) Cell surface protein partially restores morphology, adhesiveness and contact inhibition of movement to transformed fibroblast. Proc. Nat. Acad. Sci. U.S.A. 73, 1217-1221.

YAMADA, K.M., OLDEN, K. & HAHN, L-H.E. (1980) Cell surface protein and cell interactions. In: S. Subtelny (Ed.), The Cell Surface: Mediator of Developmental Processes, Academic Press, New York, (in press).

ZARDI, L., SIRI, A., CARNEMOLLA, B., SANTI, L., GARDNER, W.D. & HOCH, S.O. (1979) Fibronectin: a chromatin-associated protein? Cell 18, 649-657.

ZETTER, B.R. & MARTIN, G.R. (1978) Expression of a high molecular weight cell surface glycoprotein (LETS protein) by preimplantation mouse embryos and teratocarcinoma stem cells. Proc. Nat. Acad. Sci. U.S.A. 75, 2324-2328.

ZETTER, B.R., MARTIN, G.R., BIRDWELL, C.R. & GOSPODAROWICZ, D. (1978) Role of the high-molecular weight glycoprotein in cellular morphology, adhesion and differentiation. Ann. N.Y. Acad. Sci. 312, 299-316.

ZEUTHEN, J., NØRGAARD, J.O.R., AVNER, P., FELLOUS, M., WARTIOVAARA, J., VAHERI, A., ROSEN, A. & GIOVANELLA, B.C. (1980) Characterization of a human ovarian teratocarcinoma-derived cell line. Int. J. Cancer 25, 19-32.

© 1980 Elsevier/North-Holland Biomedical Press
Development in Mammals, Vol. 4, M.H. Johnson, editor

ANTIBODIES AS TOOLS TO INTERFERE WITH
DEVELOPMENTAL PROCESSES

Ch. Babinet and

H. Condamine

Service de Genetique Cellulaire de

l'Institut Pasteur et du College de France

25, rue du Docteur Roux

75015 - PARIS, FRANCE

The strategy of using antibodies to gain insights into
developmental problems can be formulated in the following general
way. Consider a molecular structure present in cells undergoing
some process of differentiation and/or morphogenesis. Provided
this structure is immunogenic, one may raise antibodies against
it. The question then is: what effect does the binding of the
immunogenic structure with its antibodies have upon the developmental sequence under study? If the antigen is borne on the cell
surface, the problem is solved simply by adding the antibodies
to the medium in which the cells are placed. More sophisticated
techniques will be required if the antigen is intracellular.

It is the purpose of this paper to review the main results
which this particular use of antibodies has made possible during
recent years. It is worth remembering, however, that considerable
numbers of observations along these lines have accumulated over
many years. As early as 1955, Nace was reviewing a vast literature on the same subject, with a quotation from Sigurdson which
shows that in 1940 the strategy outlined above had been defined
clearly, its goal being to determine "what components of the cell
are so vitally important that their blocking by antibodies is
incompatible with the growth and proliferation of the cell". By
the time Nace was writing his review, two important points had
been recognised. First, it had been shown in a number of
organisms that cellular antigens fall into two classes, some being
present during particular phases of ontogeny only, while others
can be demonstrated throughout development or nearly so (this
restriction applying to the earliest stages of embryogenesis).
Equally important was the fact that, in addition to their temporal
specificity, antigens of the first class could also have a specific tissue distribution (for a more recent review, see Croisille,
1970). This went in line with the concept of cell differentiation
resulting from a particular set of proteins being produced by the
cell (Monod & Jacob, 1961) and led rather directly to the
definition of differentiation antigens (Boyse & Old, 1969; see
section 6). Secondly, there were known cases in which antibodies
had important specific effects on development. Two examples,
involving unrelated systems, will illustrate this point: (a) In a
study by Spiegel (1955) antisera raised against isolated sponge

cells would inhibit the aggregation of cells from the same sponge species, but not from different species, and (b) Chick embryos grown in the presence of anti-heart antibodies had the development of their heart inhibited specifically (Ebert, 1950; Ebert *et al*, 1955).

Basically, two different approaches can be taken to make use of the immunological tool in the study of ontogeny and differentiation. The choice between the two does not in general depend on the experimentalist, rather it is imposed by the particular features of the system under study: (a) When a molecule (protein, glycoprotein, glycolipid, etc.) known to be present in cells at certain stages of a developmental sequence has been obtained in a purified form, antibodies can be raised against it and used to explore the role of that molecule specifically. In its most recent version, this approach takes advantage of monoclonal antibodies, the production of which has been made possible by the techniques of Köhler and Milstein (1975); (b) Alternatively, antibodies can be raised blindly against whole cells from a cell line established *in vitro* or from an embryonic tissue. If these antibodies turn out to have any specific effect upon the differentiation of the immunizing cells, one can look for a cell component which inhibits the effect of antibodies. Ultimately, one should be able to isolate one or several antigens, the function of which has been revealed by the antibodies. In contrast to the former strategy, this latter one is always available in principle. A majority of the experiments reported here have made use of it. It should be remembered, however, that the antisera made in this case are always complex, which may be a source of ambiguity in the interpretation of results.

The field opened by this methodology is practically unlimited. This review will be limited to the description of experiments in which the fate of whole embryos or embryonic cells was examined when grown *in vitro* in the presence of a given antiserum. Species very distant from each other from an evolutionary point of view have been used for this purpose. It is hoped that putting together observations made on very different systems but using the same approach will be of interest, even though it was not possible to give all the details on each separate system.

The limits thus placed on the scope of this review will
result in the exclusion of related topics of interest, such as
(i) the effect of antibodies on membrane antigen expression, that
is, the phenomenon of antigenic modulation (see for example,
Cohen & Liang, 1979), (ii) the use of antibodies to study the
various steps of fertilisation (see Meizel, 1978; Yanagimachi
et al, 1976), (iii) the use of antibodies as teratogenic agents
acting upon the *in vivo* development of the mammalian embryo (see
references in Leung, 1977), and (iv) the use of antibodies as
probes in the study of host-parasite relationships (references in
Pool, 1979). Moreover, the use of lectins, which specifically
recognize glycosidic residues borne by cell surface molecules,
will not be considered here.

1. THE IDENTIFICATION OF CELL COMPONENTS INVOLVED IN CELL-CELL ADHESION MECHANISMS

Experiments performed on early mouse embryos, chick embryonic
neural cells, sea urchin embryos and slime moulds will be reviewed
in this section. They exemplify the "blind" methodology outlined
in the introduction. Xenogeneic antisera raised against whole
cells were found to inhibit the aggregation of immunising cells,
thus leading to the identification of "cell adhesion molecules".

1.1 THE EFFECT OF RABBIT ANTI-MOUSE EMBRYONAL CARCINOMA ANTIBODIES ON THE PREIMPLANTATION MOUSE EMBRYO

In order to look at possible involvement of cell surface
structures in the first events of differentiation and morphogenesis of the preimplantation mouse embryo, experiments have been
performed in which the preimplantation stages were cultured *in
vitro* in the presence of antibodies against embryonal carcinoma
(EC) cells (Babinet et al, 1977; Kemler et al, 1977; Johnson et al,
1979). These cells resemble, from several points of view, a sub-
population(s) of the early embryonic cells (Graham, 1977; Jacob,
1978). In particular, their surfaces bear antigenic structures
which are also present on cleaving eggs and absent from differentiated cells. Two-cell cleavage stage embryos were put in
culture in the presence of monovalent Fab fragments prepared from
antibodies against F9 cells, an embryonal carcinoma cell line,
raised in a rabbit (Babinet et al, 1977; Kemler et al, 1977).

Under these conditions, blastomeres continued their divisions, but instead of following the normal morphogenetic pathway, i.e. compaction and divergence of two classes of cells (inner cell mass (ICM) and trophectoderm) to give a blastocyst, they formed a grape-like structure with loosely attached cells. This effect could be reversed: when grape-like structures were washed free of antibody and recultured in fresh medium, they would recompact and form a blastocyst which could eventually develop into a normal embryo and, upon reimplantation into a foster mother, would yield a mouse. This effect seems to be specific, as several other antisera containing antibodies directed against surface structures present on EC cells, blastomeres and differentiated cells, did not prevent the formation of a blastocyst from the 2-cell stage. Using antibodies directed against LS 5770, another EC line, Johnson et al, (1979) obtained similar results, namely compaction and blastocyst formation were inhibited. The effect of prolonged decompaction was studied more particularly by these authors: 2-cell embryos were cultured for increasing periods of time in the presence of antibodies and subsequently rinsed and recultured in fresh medium. It was found that compaction could be delayed up to 20 hours beyond its normal time without impairing the formation of a blastocyst. When compaction was delayed further, by leaving embryos in the presence of antibodies for longer periods of time, the proportion of normal blastocysts fell.

The appearance of the embryos cultured in the presence of rabbit anti-EC antibodies suggests that the primary effect of this treatment is to impair the normal relations between the cells in the cleaving embryo. Flattening of cells upon each other at compaction is prevented and cells appear only loosely attached to each other. It is known that the process of compaction is accompanied by the formation of different types of intercellular junctions, resulting around the 16-32 cell stage in the formation of a cohesive outer layer of cells surrounding the ICM cells. It was thus tempting to assume that the anti-EC antibodies might have an effect on the formation of these junctions. Indeed, it was shown with the use of electron microscopy that grape-like structures were deprived of any junctional complexes between blastomeres (Johnson et al, 1979). This point was also investigated

using embryonal carcinoma cells. When put in culture, these cells
have a strong tendency to aggregate and exhibit extensive
junctional complexes, including both gap and tight junctions, as
is the case in the late morula. When anti-F9 Fab were added to
the culture at the time of plating, cells would not aggregate and
remained isolated. When Fab was added after the cells had been
plated and allowed to form aggregates, a rapid decompaction effect
was evident within one hour, with each cell rounding up and
becoming isolated. During this process, the tight and gap
junctions were shown to be progressively destroyed. After longer
periods of incubation (30 hours or more), the junctions dis-
appeared completely (Dunia et al, 1979). It should be noted
that, as was the case with the preimplantation embryo, EC cell
division was not prevented in this experiment and the effect of
antibodies was reversible. Using the same system, it was shown
further that anti-F9 Fab inhibited metabolic cooperation between
EC cells (Nicolas et al, 1980). In all these experiments, controls were performed indicating that the effects observed were
the result of interaction between anti-EC antibodies and specific
receptor sites and could not be interpreted as due to some non-
specific coating of the cell surfaces by the antibodies.

Using the inhibition of decompaction by the Fab as a test,
it was possible to purify a glycoprotein from membranes of EC
cells which competes for the effect of anti F9 antibodies on cell
aggregation. When added to early uncompacted morulae grown in
the presence of these antibodies, development proceeds to the
blastocyst stage normally. Thus, this protein which has a molecular weight of 84,000 daltons, seems to be the target (or one
of the targets) implicated in the effects generated by rabbit
anti-F9 antibodies (Hyafil et al, 1980). These experiments show
that this protein is present at the surface of both cleaving
blastomeres and EC cells. No data are, as yet, available about
its distribution in later stages of the embryo, but one could
speculate about the possible specificity of such molecules
according to the stage of development. From this point of view,
it is interesting to note that the decompacting activity of the
rabbit anti-F9 antiserum could be absorbed with cells corresponding to early developmental stages such as EC cells or

parietal yolk sac cells, but not with more differentiated cells (Nicolas *et al*, 1980).

As postulated several years ago, (Mintz, 1965; Tarkowski & Wrobleska, 1967), it is generally assumed that cell fate in the preimplantation mouse embryo is determined according to position, cells on the outside of the embryo giving rise to the trophecto-derm, while those inside give rise to the inner cell mass. According to this hypothesis, there must be some clues which indicate to a cell whether it is in an "inside" or in an "out-side" position. It is known that different types of junctions between cells appear around the 8-16 cell stage and develop thereafter. These junctions which differ between outside cells and inside cells might provide positional clues by virtue of microenvironmental differences (Ducibella, 1977). As described above, the effect of the anti-EC antisera is to disturb normal relationships between cells in the embryo; all cells appear round and loosely attached to each other as is the case in the normal 8-cell embryo prior to compaction. It would seem that each cell finds itself in a situation equivalent to "outside position". In particular, each of them is sensitive to complement dependent immunolysis (Solter & Knowles, 1975; Handyside, 1978), whereas the same treatment applied to control compact morulae always yields a core of live inside cells (Johnson *et al*, 1979). In view of this result, the fact that the effect of anti-EC anti-bodies is reversible might indicate that the fate of cells remains labile for a certain period of time. An alternative explanation is not excluded, however, namely that despite outside conditions differentiation proceeds along two different lines (ICM and Trophectoderm) and that, upon removal of antibodies, sorting out of cells takes place and normal development resumes. From this point of view, it would be of interest to look at the status of each individual cell in the grape-like structure obtained in the presence of antibodies. Were the latter explanation to hold true, the inside-outside hypothesis would probably need revision.

1.2   THE IDENTIFICATION OF CELL ADHESION MOLECULES
("CAM") FROM CHICK EMBRYONIC NEURAL CELLS

Specific antibodies have been used to identify a cell surface protein involved in the mechanisms of adhesion among neural cells of the chick (Rutishauser & Edelman, 1978). The immunological reagent was produced by hyperimmunising rabbits with 10-day old embryonic retinal cells which had been cultured in suspension for 20 hours (Brackenbury *et al*, 1977). Monovalent Fab fragments were prepared from the immunoglobulin fraction of the resulting antiserum.

When 10-day old retinas are dissociated by mild trypsinization and pipetting and kept in suspension at 37°C, they readily form aggregates whose appearance can be followed quantitatively. When immune monovalent Fab' fragments are added to the culture medium, however, this aggregation process is inhibited to an extent which varies considerably with antibodies from different rabbits but may be quite high. It is worth noting that divalent immunoglobulins, from which the monovalent Fab fragments are derived, far from inhibiting cellular aggregation under these conditions, promote retinal cell agglutination. The Fab' dependant aggregation inhibition provides in turn a way to look for membrane-derived molecules which interact with the Fab' fragments and are able as such to reverse the inhibition phenomenon. Using this strategy, Thiery *et al*, (1977) have been able to purify and characterize a cell adhesion molecule ("CAM") which is involved in the aggregation of various neural cells, including cells from retina, brain and spinal ganglion, but not in the adhesion of liver cells. Under denaturing conditions, the CAM material appears to have a molecular weight of 140 000. It is readily detected by an immunoperoxidase procedure on the surface of retinal cells (Rutishauser *et al*, 1978a). When whole retinas are kept in culture, they shed CAM into the medium, both in its intact form and split in smaller fragments which retain the antigenic properties of the original CAM.

Numerous problems remain to be solved concerning the structure and function of CAM. While there are a few indications that it might exist in an oligomeric state in the membrane (Thiery *et al*, 1977), the precise structure is unknown. That CAM is playing

a specific role in the adhesion of neural cells is shown by the
fact that various molecules which bind to the cell surface such
as succinyl-Concanavalin A, or Fab' fragments which bind to cell
surface carbohydrates (Sela & Edelman, 1977), do not inhibit the
adhesion process (Brackenbury et al, 1977). This, however, does
not say anything about the mechanism by which CAM promotes aggregation among retinal cells. Although speculative models have
been presented (Rutishauser et al, 1976), it is not known whether
CAM is involved directly as a ligand in cell-cell bonds or exerts
some indirect control on their formation. Just as important and
obscure at the present time is the problem of what role this
molecule is playing during development of neural tissues. There
is evidence, stemming from antibody absorption experiments, that
CAM is present in higher concentration on the surface of 8-day
old embryonic cells than on 14-day old cells (Rutishauser et al,
1978a). This correlates with the higher ability of 8-day old
cells to aggregate in vitro as opposed to 14-day cells. Immunoperoxidase staining observed by electron microscopy reveals that
CAM is especially abundant on neuronal processes. When 8-day
old retinal cell aggregates are kept in culture, neurites and
synapses are formed which mimic the differentiation of a normal
retina. The presence of anti-CAM Fab' fragments in the culture
medium, however, inhibits this process. It can be shown in particular that a sorting out between cell bodies and neurites which
normally occurs in such aggregates is prevented when anti-CAM
Fab' fragments are added. While cellular differentiation seems
to take place in the retinal aggregates, the spatial distribution
of the differentiated components is very much altered (Rutishauser
et al, 1978b). In other in vitro experiments, an effect of anti-
CAM Fab' on growth of spinal ganglia has been observed (Rutishauser
et al, 1978b). When the latter are placed in an appropriate
culture medium they become surrounded by a crown of thick
fascicles, formed each by the association of numerous thin
neurites. Again, anti-CAM Fab' fragments inhibit the formation
of these processes, the neurites, whose growth is not affected,
remaining isolated. It is thus remarkable that in all instances
which have been studied so far, CAM would appear to be involved
not in the processes of neural cell growth and differentiation,

but rather in the geometrical arrangement of such differentiated components. An interesting possibility might be that
CAM is involved in the association and directed growth of
neurites which form nerve trunks in the chick embryo (Rutishauser
et al, 1978b).

Another example of the immunological approach for the study
of cell-cell adhesions mechanisms is provided by the work of
Urushihara *et al* (1979). This stems from a previous demonstration by Takeishi (1977) that two mechanisms, one Ca^{++}-
dependent and one Ca^{++}-independent are responsible for the mutual
adhesion of Chinese hamster V79 fibroblastic cells. Experimental
conditions have been found which allow the preparation of cells
with both, one or neither mechanisms intact. Urushihara *et al*,
have prepared the Fab fragment from the immunoglobulin G fraction
of a rabbit antiserum raised against V79 cells which retain both
mechanisms in a functional state. When V79 cells with the Ca^{2+}-
independent adhesion mechanism intact are allowed to reaggregate
in the presence of such Fab, inhibition of aggregation occurs.
Trypsinization will release a material from the surface of V79
cells which blocks this inhibition, so that the same assay which
was used by Edelman and co-workers for the purification of CAM
is usable in this system. It should be noted that, in contrast,
the Ca^{2+}-dependent mechanism of adhesion is not blocked by the
anti-V79 Fab fragments. Also, formally similar mechanisms have
been found to be responsible for the adhesion of neural retinal
cells from 7-day chick embryo (Takeichi *et al*, 1979). The sensitivity to trypsin, in the presence or absence of Ca^{2+}, is very
similar for both mechanisms, as they are present in V79 and in
chick retina cells. Although this might suggest some structural
relationship between chick and hamster molecules, the Ca^{2+}-
independent aggregation of chick retinal cells is not inhibited
by anti-V79 Fab fragments. A species specificity (chick versus
hamster) may be superimposed in this case on a tissue specificity
(neural versus fibroblastic).

1.3 AGGREGATION FACTORS IN SEA-URCHIN EMBRYOS

Regional differentiation in the sea urchin embryo is largely
determined by the distribution of animal-vegetal components (see
for example, Hörstadius, 1973). Dissociated cells from the

blastula or mid-gastrula are able to form aggregates which may
eventually give rise to more or less normal Pluteus larvae and
this process has been shown to be species-specific (Giudice,
1962; Giudice & Mutolo, 1970; Spiegel & Spiegel, 1975). The
importance of the cell surface membrane in the development of the
early sea-urchin embryo has long been postulated (see for example
Gustafson & Wolpert, 1967). To look more specifically at cell
adhesion, a quantitative assay was devised by McClay and Hausman
(1975): early embryos were dissociated in Calcium-Magnesium-free
sea water, and the isolated cells allowed to reform aggregates,
which grew in size with time. It was shown that these aggregates,
left in suspension, could differentiate up to the Pluteus stage.
To test the ability of embryonic cells to adhere to aggregates,
the former were labeled with H^3 leucine and added to the aggregates. After a given time the proportion of cells which had
adhered to the aggregate was obtained by measuring the radio-
activity of the latter. The adhesive properties of embryonic
cells from two different sea-urchin species *Lytechinus variegatus*
(LL) and *Tripneustes eschulentes* (TT) and their reciprocal hybrids
($♀$ L × $♂$ T = LT ; $♀$ T × $♂$ L = TL) were examined by this method.

It was shown that homo-specific adhesion (*LL + LL* or *TT + TT*)
was much more efficient than heterospecific adhesion (*LL + TT*)
whether blastula or gastrula cells were considered. In contrast,
when hybrid cells, *TL* for example, were used before the gastrula
stage, they would adhere only to maternal homospecific cells (*TT*)
but not to the paternal ones (*LL*). If however, *TL* cells were from
the gastrula stage or older, they would exhibit the ability to
adhere to cells of the paternal genotype, i.e. *LL* cells, indicating that *L*-specific determinants started being synthesized under
the direction of the paternal genome around gastrulation.

Subsequently, an antiserum (*LL* antiserum) was prepared
against *Lytechinus* midgastrula membranes (McClay *et al*, 1977).
In indirect immunofluorescence experiments, this serum was shown
to label *LL* and *LT* cells of both pregastrula and postgastrula
stages, whereas *TL* cells were labeled only if taken from postgastrula stages. Monovalent Fab fragments were prepared from this
serum and pre-absorbed on *TT* cells (*LL*-Fab). When *LL* pregastrula
cells were preincubated with *LL*-Fab, their collection by *LL*

aggregates was inhibited. Preabsorption of LL-Fab by LL cells would completely prevent the inhibition. It should be noted that if whole LL-serum was used in place of LL-Fab, LL collection of cells was stimulated (McClay et al, 1977).

This property of the whole serum was used to further characterize different classes of surface determinants according to the stage of development (McClay & Chambers, 1978; for a review, see McClay, 1979). It was first demonstrated that the LL antiserum promotes agglutination of LL cells whereas pre-immune serum does not. Then using selective preabsorptions with cells from embryos of different stages followed by an aggluti-nation test of the absorbed serum, it was possible to define at least four different classes of antigens: one present from the beginning of development, a second one appearing around gastru-lation and two others specific for the ectoderm and the endoderm of the Pluteus larva respectively (McClay & Marchase, 1979). The exact nature of the antigenic site(s) recognized by the antiserum, and how they are acting in the process of specific cell adhesion remains to be established.

The problem of identifying surface sites possibly involved in specific cellular adhesion of the sea-urchin larva was approached in another recent study: monovalent Fab fragments were prepared against purified membranes from Paracentrotus lividus late blastulae (Noll et al, 1979). When added to a culture of dis-sociated blastula cells these Fab fragments would prevent re-aggregation of the cells. Moreover, they were able to promote the dissociation into single cells of either blastula or gastrula embryos. In view of the results presented above (McClay & Chambers, 1978), this observation probably means that the anti-bodies recognize more than one site implicated in specific cell-cell interaction. It was shown also that inhibition of reaggregation can be reversed if components extracted from puri-fied membranes by butanol are added together with the Fab frag-ments. Interestingly, the same components are able to enhance the reaggregation of dissociated T blastula cells by themselves. This observation prompted the authors to look for a direct action of these butanol extracted membrane components. Isolated blastula cells were treated with butanol; after this treatment, cells were

still viable but became unable to reaggregate. Restoration of
aggregation was obtained by addition of butanol extracts to the
culture. Thus a component is present in the blastula cell membrane, which can be extracted by butanol and is specifically
involved in cell-cell interactions. There are indications that
this component is of proteic nature. The fact that it is obtained
in a soluble form in the butanol extracts opens the way to its
purification.

It should be added that *Paracentrotus* cells, previously
treated with butanol, are induced to reaggregate among themselves
when treated by the butanol extract of another sea-urchin species,
Arbacia. Since the process of cellular adhesion has been shown
to be species specific (McClay *et al*, 1977), this observation
implies that the phenotype of the cells, as regards their specific adhesion properties, has been changed. These findings which
point to some proteins being easily removed from the surface of
the cells with concomitant change of surface specific properties,
can be compared with similar observations in other systems: in
the sponges, new species-specificity could be induced by transfer
of a species-specific aggregation factor (a glycoprotein) to
factor depleted cells (Muller *et al*, 1978a,b). Similarly,
addition of LETS protein to transformed chicken fibroblasts from
the outside would restore a phenotype of untransformed fibroblast
(Yamada, 1978).

1.4 AGGREGATION FACTORS IN DICTYOSTELIUM

The cellular slime moulds are presently the organisms in
which the use of antibodies as probes for specific molecules
involved in morphogenetic events has been the most fruitful.
The life cycle of this organism is well known (see Loomis, 1975).
When grown in the presence of bacteria as a source of food, the
amoebae will divide by fission and stay as isolated cells
("vegetative" or "growth phase" cells). When the medium is
depleted of bacteria, isolated cells undergo a complex process
of migration (see Newell, 1977) mediated by a chemical attractant,
originally called acrasin (Bonner, 1947). During this phase,
cells acquire new properties which render them mutually cohesive
("aggregation competent" cells). Thus aggregates ("slugs") are
formed which will ultimately differentiate into a structure

consisting of a mature fruiting body, supported by a stalk, the whole process taking around 24 hours. Cell aggregation which follows starvation, is a highly coordinated phenomenon and is accompanied by drastic changes in cell behaviour and properties (reviewed by Loomis, 1975). In addition, it may be triggered under very precise and reproducible conditions. It is therefore a particularly suitable system to study cellular interactions during a developmental process. First attempts to look for molecules specifically involved in cell adhesion were made with the species *Dictyostelium discoideum*.

Two types of contacts are made by the cells: end-to-end and side-by-side contacts (Beug et al, 1970). The latter are inhibited by EDTA (Gerish, 1961) whereas the former are not. Growth phase cells make only side-by-side contacts, aggregation competent cells make both kinds (Beug et al, 1970). Quantitation of aggregation was obtained by comparing light scattering of the suspension under test with that of a single cell suspension.

Monovalent antibodies, prepared by immunization against particulate fraction of aggregation-competent cells, completely inhibited both side-by-side and end-to-end contacts (Beug et al, 1970). Absorption experiments with either vegetative or aggregation competent cells indicated that each type of contact is blocked by Fab of different specificity (Beug et al, 1973a), the target antigens being called respectively contact site B (side-by-side, "cs B") and contact site A (end-to-end, "cs A"). Furthermore it could be shown that cs-A were not present on growth-phase cells but rather appeared during the phase where the cell acquires the aggregation competence, thus being developmentally regulated. When Fab directed against sugar moieties of the cell surface (and shown to bind to the latter) were used, no inhibition of aggregation was observed suggesting that this inhibition was not merely due to some non-specific "coating" of the cell surface. In a further study, where labeled Fab was used, it was estimated that around 10^5 Fab molecules per cell were sufficient to completely block contact sites A, whereas up to eight times as much Fab directed against sugar moieties on the cell surface failed to impair cell aggregation (Beug et al, 1973b). These observations also supported the idea that aggregation inhibition was due to a masking by Fab of specific targets at the surface of the cell, the presumed contact site A. The latter can be assayed by absorption of aggregation-inhibiting activity and therefore a quantitative assay is available for their purification: any fraction to be tested is incubated with a known quantity of blocking Fab, and if it contains cs-A, a decrease in adhesion blocking activity will result. Using this test, it was possible to purify a protein which co-fractionated with cs-A activity. This protein could be obtained as a single band in polyacrylamide gel electrophoresis; it is a glycoprotein which binds concanavalin A with an apparent molecular weight of 80,000(Huesgen & Gerish, 1975; Muller & Gerish, 1978). The purified protein contained all the determinants involved in the blocking effect: its addition to a polyspecific adhesion-blocking Fab preparation resulted in a complete loss of inhibition of adhesion.

Plasma membranes of *D. discoideum* contain a number of glyco-proteins and lectin-like proteins which have been analysed in relation to stages of development (Geltosky *et al*, 1976; Reithermann *et al*, 1975; Parish *et al*, 1978). Two of these, the lectin-like discoidins I and II were purified and shown to have related, although not identical, properties (Frazier *et al*, 1975). *In vitro* they are synthesized by aggregation-competent cells and not by vegetative cells (Rosen *et al*, 1973, Ma & Firtel, 1978) and they are present at the surface of cells (Chang *et al*, 1975). They have been purified and shown to be tetramers, their subunits having molecular weights of 24 000 and 26 000 respec-tively (Frazier *et al*, 1975; Ma & Firtel, 1978). A mutant has been isolated which lacks one functional discoidin, the other being normal. This mutant is unable to aggregate (Ray *et al*, 1979). Taken together these properties suggest strongly that discoidins also play a role in cellular aggregation. Another glycoprotein, gp 150, was shown to be developmentally regulated; it is not present in vegetative cells but may be detected in large amounts in aggregation competent cells (Geltosky *et al*, 1976). On these grounds, Geltosky *et al* (1979) purified it and raised antibodies against it. Immunoprecipitation of the total concanavalin A binding protein fraction by this antiserum resulted in a marked enrichment in Gp 150 protein, thus assessing the

specificity of the antibodies. Furthermore, monovalent Fab'
fragments prepared from the anti-Gp 150 antiserum were able to
completely inhibit the adhesion of aggregation competent cells.
The apparent molecular weight of this protein was determined as
being around 150,000. Although it was demonstrated by specific
immunoprecipitation to be different from cs-A, it shares several
properties with it: it is a glycoprotein, its synthesis is stage
specific, being present only in aggregation competent cells and
monovalent antibodies raised against purified gp 150 completely
block cell adhesion. Similar results have been obtained with a
related slime mould species, *Polysphondylium pallidum* (see Rosen
et al, 1976, 1977; Bozzero & Gerish, 1978; Steineman *et al,* 1979).
At the present time, four different components have been shown
to be implicated in the specific adhesion of differentiating
Dictyostelium cells: these are cs-A, cs-B, gp 150 and discoidin.

In view of these results, two questions may be asked:
1) Are these components independently involved in the process of
cell adhesion? 2) What are the molecular mechanisms acting at
the cell surface during this process? Clearly cs-A and cs-B
are independent of each other: in the first place, they mediate
different types of cell contact, side-by-side and end-to-end
(Beug *et al*,1973), the latter being uncoupled by EDTA (Gerish,
1961); secondly, they are blocked by Fab of different specificities (Beug *et al,* 1970). Also, gp 150 is different from cs-A,
the compelling evidence, besides different molecular weights,
being that Fab against purified gp 150 does not immunoprecipitate
the cs-A protein (Geltosky *et al,* 1979). Finally, both lectin-
like biological properties (hemagglutination of sheep red blood
cells) and molecular properties differentiate discoidin from cs-A
and gp 150.

Taken together, these facts suggest strongly that several
independent molecular systems are acting in the developmentally
regulated cellular adhesion of *Dictyostelium*. From this point of
view, it might be interesting to look more closely at the effect
of the different Fab fractions upon already formed aggregates,
taken at different times of normal development. The actual
molecular mechanisms in which these different proteins are
involved remain obscure, although different models have been

proposed (see for example, Ray *et al*, 1979; Muller & Gerish, 1978). In essence, these are of three kinds and postulate cell-cell recognition through (a) two different but complementary structures, or (b) through transmembrane homodimer formation, or (c) transmembrane association of two structures through an intermediate ligand.

In the case of cs-A and gp 150, a single glycoprotein was able to remove all the Fab blocking activity, which would be in favour of model b (see Muller & Gerish, 1978). As regards discoidin, model a has been favoured (Ray *et al*, 1979) on the grounds that HJR 1, a mutant lacking a functional discoidin and unable to aggregate, could be rescued completely by normal cells. The interpretation was that HJR had a normal receptor to dis-coidin which could interact with discoidin of the normal cells. Thus it could be that not only different sites are implicated in cell-cell adhesion but that they are acting through different molecular mechanisms.

1.5 CONCLUSIONS

A rather uniform scenario is at work in the four widely different systems which have been reviewed so far. Antibodies raised against whole cells or membrane cells have a dissociating effect upon some specific target (either the immunizing cells themselves or related cells). This provides a way of assaying molecules able to block this effect. A number of proteins or glycoproteins have been obtained in this way, which are clearly responsible for cell-cell adhesion mechanisms. Many problems, however, remain to be solved, including: (a) the detailed structure of cell adhesion molecules. It is too early yet to say that these molecules all belong to a homogeneous class with common structural features; (b) their degree of specificity (both species and tissue specificity); and (c) the chronology of their expression during embryogenesis and differentiation. Ultimately, this information should converge to elucidate the nature of adhesion mechanisms of which these molecules are an essential components.

2. THE ANALYSIS OF H-Y ANTIGEN DIRECTED GONADAL MORPHOGENESIS

The H-Y antigen offers a unique situation in which a molecule, detected immunologically, appears to be implicated in a
sequence of both morphogenetic and differentiative events.
Various aspects of the H-Y antigen story have been reviewed
recently (Wachtel, 1979; Wachtel & Koo, 1980). Here the *in vitro*
experiments dealing with gonadal organogenesis and the main
problems which they raise will be considered.

H-Y antigen was first detected in skin graft rejection
experiments where it was observed that female mice would reject
skin grafts from males of the same inbred stain, while males are
tolerant to female grafts(Eichwald & Silsmer,1955). Syngeneic
anti-H-Y antisera were first obtained by Goldberg *et al*, 1971,
and appear to be cytotoxic on various target cells, including
sperm and male epidermal cells. This finding allowed in turn the
performance of indirect absorption tests to detect the presence
of H-Y antigen on any tissue. Thus, cells from a given tissue
are incubated with anti-H-Y antiserum, the residual cytotoxic
activity of which is determined after absorption. The same test
can be used as an assay for soluble H-Y antigen, which has thus
been shown to be secreted in epididymal fluid (Muller *et al*, 1978)
and in the culture medium of the human "Daudi" cell line (Koo,
quoted in Ohno *et al*, 1979; see also Beutler *et al*, 1978; Fellous
et al, 1980).

It has been recognized that gonadal morphogenesis and differentiation in mammals is correlated with the absence or presence
of H-Y antigen, rather than with an XX or XY karyotype (for some
exceptions to this rule, see Wachtel, 1979). That is, an
embryonic gonad will develop into a testis irrespective of its
chromosomal constitution provided its cells are coated with H-Y
antigen. An example is given by the *Sxr* mutant in the mouse; mice
with the XX karyotype and heterozygous for *Sxr* have (sterile)
testes rather than ovaries and these testes are positive for the
H-Y antigen (Bennett *et al*, 1977). Conversely, gonadal cells
which lack H-Y on their surface membrane will organize themselves
into an ovary, even if they possess an XY karyotype, a situation
illustrated by the XY females of the wood lemming (Wachtel *et al*,

1976). If gonadal organogenesis is primarily under the control of H-Y antigen, one should be able to influence the pattern of histiotypic reaggregation exhibited by gonadal cells which have been dissociated and allowed to reassociate according to the technique of Moscona (1957, 1961). H-Y$^+$ cells might organize themselves into follicle-like (ovarian) structures when their H-Y antigen is blocked by anti-H-Y antibodies. H-Y$^-$ cells might be induced to form structures reminiscent of seminiferous tubules, provided they can be coated with exogenous H-Y molecules. Indeed, the latter possibility is offered by the finding that an H-Y receptor is demonstrable at the surface of H-Y ovarian cells (Müller *et al*, 1978a).

Following this rationale, experiments have been performed by Ohno *et al* (1978, 1979a,b), using newborn mouse gonads and by Zenzes *et al* (1978a,b) using rat gonads. Both experimental sets have yielded essentially the same results. The reason for using newborn material of course is that at that stage some developmental lability might still be expected and could be demonstrable under proper experimental condition. A newborn mouse testis, however, is a rather complex structure with already well identifiable seminiferous tubules separated by interstitial tissue. Seminiferous tubules are made of Sertoli cells surrounding primordial gonocytes in the lumen. The interstitial tissue comprises several cell types, fibroblastic and Leydig cells being the most prominent (see Theiler, 1972). The newborn mouse ovary could possibly be considered a simpler structure with only a few follicles being organized, each comprising a central gonocyte surrounded by follicular cells.

Single cell suspensions are readily obtained by mild trypsinization of newborn mouse or rat testes. While primordial gonocytes are easily identified in such suspensions due to their large size and rounded shape, the identity of other cells remains largely undetermined. When incubated under conditions which allow the formation of aggregates, however, they mainly yield elongated twisted tubular structures, with gonocytes in their lumen. A study by Davis (1978) illustrates well the propensity of dissociated testicular cells to reorganize into tubular structures. How close to genuine testicular tubules these reconstituted tubes may be remains to be determined. As noted above, gonocytes are the only cells identifiable with some certainty in testis cell suspensions. It is tempting, however, to assume that the walls of the reconstituted tubules are primarily made of Sertoli cells and that this *in vitro* process is a good substitute for the *in vivo* testis organogenesis.

The fate of complexes of H-Y antigen with anti-H-Y antibody after antibody binding at the surface of testicular cells is not known in any detail. It is assumed by analogy with other systems (Cullen *et al*, 1973) that the complexes migrate towards one pole of the cell where they become internalized subsequently, thus stripping the testicular cells of their H-Y antigens. Alternately, one could assume that the binding of anti H-Y antibodies *per se* impairs whichever functions are fulfilled normally by the H-Y antigen. In either case, testis cells which have bound anti-H-Y antibodies might be expected to be H-Y$^-$ at least from a functional point of view. Indeed, when anti-H-Y antibodies are applied on testis cells in the process of reaggregation, an overall effect is noticeable, in which the percent of cylindrical structures obtained is diminished (anti-H-Y antiserum diluted 1/5 will reduce that fraction of tubule-like aggregates down to 30% of the control value obtained when no anti-H-Y antibody is present). In contrast, the fraction of aggregates structured in a folliculoid (ovarian) manner is greatly enhanced. This is an overall effect observed in long term experiments where aggregation is allowed to proceed for several hours. One cannot judge whether the anti-H-Y anti-serum has any effect upon the kinetics of initial reaggregation between testis cells. The reciprocal experiment, in which newborn rat ovarian cells are dissociated, coated with H-Y antigen and allowed to reaggregate has been done by Zenzes *et al* (1978a). The source of H-Y antigen in this case was the supernatant of new-born rat testicular cell cultures, an application of the finding that testis cells excrete H-Y into their culture medium (Muller *et al*, 1978c). Histological examination of ovarian cell aggregates obtained under control conditions (i.e. with no H-Y antigen present in the medium) reveals the formation of typical follicular structures as expected, with flat somatic cells arranged around a single large germinal cell. When H-Y is added, however, tubules analogous to those found in reaggregating testis cells are detectable, with several gonocytes present in their lumen. Under the conditions used by Zenzes *et al*, the testicular conversion of ovarian cells was incomplete, that is, follicular structures would still co-exist with tubules within the same aggregate. It should be noted that anti-H-Y antibodies prevent these tubules from appearing. Furthermore, it could be shown that an hCG surface receptor, normally present on newborn rat testis cells but not on newborn ovarian cells (Siebers *et al*, 1977) was progressively expressed during the reaggregation process. Cycloheximide added shortly after the onset of the experiment prevented the appearance of this hCG receptor (Müller *et al*, 1978b).

A somewhat different approach was taken by Nagai *et al* (1979) in which the proteins excreted by the human cell line Daudi were the source of H-Y antigen. Indifferent gonads from karyotyped XX bovine embryos of crown rump length 25-30mm were kept for up to five days under organ culture conditions, with and without Daudi-excreted material in the medium. While gonads remained indifferent in the control cultures, those gonads which had been submitted to Daudi's excretion developed tubules and the presence of *tunica albuginea* as assessed by histological examination. A check for the specificity of this effect, however, using an anti-H-Y serum, has not been reported (see Ohno *et al*, 1979b).

This set of experiments is remarkable inasmuch as it provides an *in vitro* confirmation of the model postulated for H-Y function on the basis of immunogenetic observations. Among the many experiments done so far in which immune sera have been used to interfere with differentiation processes, these are unique in that a morphogenetic event, rather than a mere specific cell adhesion process seems to be at stake. The level at which the H-Y antigen might act, however, is still to be elucidated. Apart from the fact that H-Y does not behave like an adhesion promoting molecule (since anti-H-Y antibodies do not prevent the reaggregation of testis cells), two modes of action (among others) are conceivable: (a) H-Y could trigger the synthesis by undifferentiated gonadal cells of various testis specific molecules, some of which would promote cellular arrangements leading to the appearance of tubular architectures. This triggering could be earlier than,

later than, or independent of gonadal cell adhesion. (b) H-Y
could be a member of that class of molecules, the existence of
which was postulated in (a), which direct the gonadal morphogenesis, once adhesion has taken place. The fact that an hCG
receptor, which is synthesized by newborn rat ovarian cells
treated with H-Y, is not detected when cell adhesion is prevented
(Müller et al, 1978b) may be relevant in this context. Presumably
one would need other such specific biochemical markers, which
would allow a distinction between an ovarian and a testicular
state of a gonadal aggregate, independently of histological
criterion, and thus would lend themselves to quantitative studies.

3. MISCELLANEOUS

3.1 ANTI-MOUSE EMBRYO ANTISERUM

Wiley and Calarco (1975) have immunised rabbits against mouse
blastocysts, an experimental approach seldom used. The obvious
reason for this is the vast number of embryos necessary for each
injection. In this case, a rabbit was given 200 blastocysts
weekly for four weeks, plus 800 blastocysts in a booster injection
four weeks later. The antiserum thus obtained was shown, in
indirect immunofluorescence studies, to react specifically with
preimplantation mouse embryos from the 4-cell to the 8 to 12-cell
stage. Membrane immunofluorescence diminishes steadily thereafter
on late morula and blastocyst stages. Cytotoxicity of the antiserum, as tested against preimplantation embryos, is stage-
specific also. While blastomeres from the 2-cell stage are left
intact, blastomeres from the 8-cell stage are lysed readily by
the antiserum at a concentration of 50%. Under in vitro culture
conditions in which 80% of 2-cell embryos reach the blastocyst
stage, the antiblastocyst serum inhibits blastocyst formation.
At a concentration of 4%, the serum prevents cleavage of the
embryos completely. As a control in this experiment, a rabbit
antiserum made against placentae from 16 to 18-day gestation mice
leaves the in vitro development of 2-cell embryos unaltered, even
though this antiserum can be shown to react with all preimplantation stages. The conclusion drawn by Wiley and Calarco
is that the antiblastocyst serum recognizes cell surface molecules

which are required for development to proceed. The nature of
these molecules, however, remains to be determined.

3.2 *ANTI-RAT YOLK SAC ANTISERUM*

Sheep antisera were prepared against whole yolk sacs dissected from rat term fetuses by New and Brent (1972). The effect
of immuno-globulins obtained from the antisera was checked in an
in vitro culture system in which rat embryos dissected at 9½ or
10½ days of gestation with their yolk sac kept intact, will follow
an apparently normal course of development for at least 24 hours.
Anti-yolk sac antibodies had a gross deleterious effect on the
development of the embryos in this system provided, however, that
the antibodies were bound to the outer layer (i.e. visceral endoderm) of the yolk sac. No effect was noticed when the antibodies
were injected inside the amniotic or exocoelomic cavities, thereby allowing their fixation upon the inner mesodermal component of
the yolk sac. It is assumed that outside antibodies acted by
preventing the trophic role of the yolk sac and the passage of
nutrients from the medium to the embryo. The specificity of this
effect is difficult to evaluate. It may be relevant to draw
attention to the study by Bernard *et al* (1977) which shows that
yolk sac visceral endoderm from early post-implantation mouse
embryo is loaded heavily with maternal immunoglobulins. This
would suggest that the effect observed by New and Brent is not
due to sheer coating of the yolk sac endoderm by immunoglobulins.

3.3 *THE EFFECT OF ANTI-FORSSMAN ANTIBODIES ON THE REAGGREGATION OF DISSOCIATED CELLS FROM CHICK EMBRYO MESONEPHROI*

Using immunofluorescence techniques, Szulman (1975) has
shown that anti-Forssman antibodies, obtained by immunizing
rabbits with sheep red blood cells, react with specific parts of
the mesonephroi from a 4-day incubation chick embryo. Thus, the
epithelial cells of the duct and collecting tubules are Forssman
positive, in contrast to the rest of the nephros which is Forssman
negative (the presence of a Forssman like antigenic specificity
has also been demonstrated on mouse preimplantation embryo cells:
Willison & Stern, 1978). When dissociated cells from the mesonephroi are put in culture, histotypic reaggregation occurs and
histological examination reveals the presence of reconstituted

tubules underlined with basement membranes in the aggregates.
Furthermore, a segregation between Forssman positive and Forssman
negative cells takes place, so that reconstituted tubules themselves are positive or negative each for its part.

When the reaggregation experiment is performed with anti-
Forssman antibodies present in the medium, the formation of
Forssman negative tubules remains undisturbed. Positive cells,
however, merely come together and form "spiderweb-like conglomer-
ates", which do not organize themselves into tubules. As
emphasized by Szulman, it is difficult to decide which step of
morphogenesis is impaired in the presence of anti-Forssman antibodies. Some process of cell-cell recognition apparently still
takes place since positive and negative cells sort out. It is
not clear whether *inter se* adhesion of positive cells is affected.
If it were not, the Forssman antibodies might reveal one of those
(so far) rare instances in which a surface structure is implicated
in a morphogenetic step after cell adhesion, a situation which
could be analogous to the one revealed by the anti-H-Y antibodies.

4. THE CONTROL OF CELL GROWTH AND TRANSFORMATION WITH ANTI-GANGLIOSIDE AND ANTI-LETS ANTIBODIES

The transition of a cell from the normal to the transformed
or malignant state is accompanied by several modifications in its
properties. These include, in particular, changes in substrate-
adhesion properties, new morphology, loss of contact inhibition,
new cell-surface properties. It is beyond the scope of this
review to deal with all these aspects of neoplastic transformation.

We shall concentrate on the description of experiments which
have used specific antibodies to ascertain the importance of some
cell-surface structures in the process of transformation. These
include a cell surface glycoprotein, called LETS (Mautner & Hynes,
1977) or LSP (Yamada, 1978) or fibronectin (Keski-Oja *et al*,
1976), and the family of glycolipids.

The pattern of glycolipids in animal cell membranes has been
extensively studied. It was shown that both the pattern of
synthesis and the degree of exposure of the cell-surface glycolipids varied depending on the cell cycle phase examined and on
the normal or malignant status of the cell. (For review, see
Hakomori, 1975 and Richardson *et al*, 1975; see also Critchley,

1979). Lingwood and Hakomori have suggested a role based on these observations for some glycolipids in the control of cell cycle and cell proliferation. To investigate this point, they raised antibodies against two of the cell surface glycolipids, namely N-acetyl-hematoside (GM3) and globoside (Lingwood & Hakomori, 1977), and looked for a possible effect on cell growth. Mouse 3T3 cells, hamster NIL cells and their transformed derivatives (3T3 with polyoma virus and NIL with murine sarcoma virus) were used in this study. When cells were cultured in the presence of anti-GM3 monovalent antibodies, growth rate and cell saturation density of normal cells were reduced whereas their transformed counterpart remained unaffected. In contrast, anti-globoside Fab fragment or anti-GM3 Fab absorbed with GM3-containing liposomes did not perturb the cell growth of any of the cell lines. Interestingly, an effect of anti-GM3 Fab on ganglioside metabolism could be demonstrated: when anti-GM3 Fab were added to a culture of 3T3 cells, synthesis of GM3 and GD1a gangliosides was enhanced, whereas GM2 and GM1 were unaffected, as compared to control 3T3 cells grown in the absence of antibodies.

The effect was observed only in cell culture before cell confluency. At or after confluency, GM3 and GD1a synthesis was reduced as compared to the controls. Since a similar phenomenon, namely the enhanced synthesis of certain glycolipids, is observed by the time normal cultured cells enter in contact and become contact inhibited, the authors hypothesized that anti-GM3 antibodies might in some way trigger surface changes similar to those which normally induce growth inhibition in a confluent cell culture. If so, transformed cells would be insensitive to this effect.

In a subsequent study, Lingwood et al (1978) looked more specifically at the effect of antibodies to certain gangliosides and to LETS-protein on the expression of the transformed phenotype, using a rat kidney cell line (NRK) infected with a thermosensitive mutant of Rous sarcoma virus, LA25 (NRK-LA25). They first showed that there was a lower degree of exposure of both GM3 ganglioside and LETS protein at the cell-surface in the transformed state, i.e. at the permissive temperature (32°C), than was observed at the non-permissive temperature (39°C). Accordingly,

GM3 and LETS are, at least in this system, transformation-sensitive surface structures. NRK LA25 cells, when grown at the permissive temperature, assume a round shape and may grow in 0.3% agar, whereas at the non-permissive temperature, they assume a fibroblast-like (contact-inhibited) appearance and cannot grow in agar. If however NRK/LA25 cells were grown at 39°C in the presence of either anti-LETS or anti-GM3 Fab fragments and then shifted to 32°C, they retained the fibroblastic shape and were unable to grow in agar. In contrast, anti-globoside or anti-GM1 ganglioside Fab were without effect on these phenotypic modifications linked to the transformed state. Similar results were obtained in another system of transformed cells, i.e. murine sarcoma virus infected 3T3 cells (Lingwood et al, 1978).

The way in which antibodies to LETS or ganglioside act in preventing the phenotypic expression of the transformed state is not understood but these experiments point to their importance in the expression of transformation. It should be noted that LETS protein has been recognized in a number of studies as being depleted or dismissed when cells become transformed (Hynes, 1976; Yamada & Pastan, 1976). Reconstitution experiments have been conducted in which addition of the protein to transformed cells would permit restoration of the properties of normal non-transformed cells (see for example, Yamada et al, 1976 - also Wartiovaara, this volume).

Conversely, anti-LETS affinity-purified antibodies, when added to a culture of normal chicken fibroblasts, induce a rapid change in cell morphology: cells round up and acquire a shape typical of transformed cells. This also occurs with reconstituted cells (see above). Furthermore, in this case, capping of fibronectin was observed as in normal cells. This result indicates that fibronectin exogenously added to a transformed cell assumes relations with the membrane environment similar to that in non-transformed cells (Yamada et al, 1976; Yamada, 1978).

5. NEW TECHNIQUES

5.1 *MONOCLONAL ANTIBODIES AS PROBES FOR DEVELOPMENTAL PROBLEMS*

The technique introduced by Köhler and Milstein (1975) allows production of permanent cell lines secreting an antibody of defined molecular species and specificity. Little has been published so far, dealing with the interference of such antibodies with differentiation or development sequences. A study by Dulbecco *et al* (1979b) appears to be somewhat pioneering in this respect.

These authors have examined the effect of anti-Thy-1 antibodies on the *in vitro* differentiation of a rat cell line derived from a dimethylbenzanthracene-induced mammary carcinoma. This cell line (LA7) undergoes various differentiation processes in culture, which lead to the appearance of fusiform cells and particular cell arrangements, including projections, ridges and domes (Dulbecco *et al*, 1979a). Domes, whose formation can be induced by dimethysulfoxide, are described as "blisters enclosing a pocket of liquid between the cell layer and the plastic dish". Immunofluorescence readily demonstrates the presence of Thy-1 antigen on the surface of LA7 cells. When mouse monoclonal anti-Thy-1 antisera are added to the culture medium, the LA7 cells stop producing domes, pre-formed domes disappear and dimethyl sulfoxide is no longer able to induce their appearance. Different monoclonal antibodies have different efficiencies in preventing dome formation and one antiserum has been found inactive in this respect. Antisera made against the whole cells were also found poorly active, although highly cytotoxic against LA7 cells. Two important aspects of monoclonal antibodies are exemplified in this study: (a) their potentially much higher efficiency, as compared with conventional antisera, to interfere with differentiation processes, (b) the high variability in the efficiency of independent monoclonal antisera. As stressed by Dulbecco *et al* (1979b), this approach might help to elucidate the role of different parts of an antigen in a differentiation sequence, assuming that differences in monoclonal antibody efficiency are correlated with the fact that various such antisera recognize different structures in the antigen.

*5.2 THE EFFECT OF ANTIBODIES AFTER THEY HAVE BEEN
INTRODUCED INSIDE THE CELLS*

In all instances reported so far, antibodies were simply
added to the culture medium of cells upon which their effect was
to be studied. The presence of antigenic receptors on the outer
part of a cell surface membrane is required in any study of this
type. Recent investigations, however, have shown that it may be
possible to introduce the antibodies inside the cells and follow
their actions for rather long periods of time.

In a study by Mabuchi and Okuno (1977), rabbit antibodies
were prepared against egg myosin from the starfish *Asteria
amurensis* and microinjected in one blastomere of the 2-cell
embryo. While the untreated blastomere would divide normally,
conditions could be found which blocked the mitotic activity of
the injected blastomere completely up to eight hours. Some process of nuclear division, however, was still noticeable. Rungger
et al (1979) were able to inject antibodies into the nucleus of
Xenopus oocytes, using a modification (Rungger & Türler, 1978) of
a technique established for the microinjection of DNA (Brown &
Gurdon, 1977). They could observe the inhibition of mitotic
spindle formation due to the intranuclear injection of human anti-
actin antibodies. The same antibodies remained without effect
when injected into the cytoplasm. Similarly, rabbit anti-myosin
antibodies, injected into the nucleus, had no effect. Scheer *et
al* (1979) were also able to inject rabbit anti-histone antibodies
into the nucleus of *Pleurodeles* oocytes. Retraction of chromosomal loops followed and chromosomes lost their characteristic
lampbrush morphology. All these studies demonstrate that antibodies can be injected into cellular cytoplasmic or nuclear compartments and retain their activity for several hours. They show
that the exploration of molecular functions at the intracellular
level is available thanks to this technique. The same conclusions
can be drawn from the experiments of Yamaizumi *et al* (1979).
These authors have devised a method in which antibodies are packed
within resealed erythrocyte ghosts. Virus-mediated fusion with
cultivated cells is then used to introduce the antibodies into the
cytoplasm of the cells (Yamaizumi *et al*, 1978a,b). The functional
stability of antibodies (rabbit anti-diphtheria toxin fragment A

immunoglobulin) inside Vero cells (a Monkey cell strain) was investigated. More than 50% of the initial neutralizing activity of the internalised antibodies was still present after the cells had been incubated for 20 hours. Thus two points are worth emphasizing: (a) the existence of techniques which allow the introduction of antibodies in isolated cells (microinjection) or in a whole population of cells (virus-mediated fusion), (b) the persistence of antibody activity after they have been injected for several hours at least.

6. CONCLUDING REMARKS

Each of the experiments reviewed here brings a partial answer to the question: what can we learn about morphogenetic and cell differentiation processes by adding specific antibodies to a system undergoing such a process *in vitro?* As was emphasized in the Introduction, this question has been asked and experimentally tried for a long time. It is but one question, however, among those which have to do with development and can be approached with immunological, or more specifically immunogenetical, tools (see Bennett *et al,* 1971; Boyse, 1979). It can be reframed in a slightly more technical way: what is the function of a differen- tiation antigen? Although strictly speaking this notion merely has to do with the morphology of cell surfaces and refers to "antigenic variation within the species, between different cell types in a single individual" (Boyse & Old, 1969), it brings with it a functional connotation of which the coiners of the term were fully aware: 'Antigens of this kind constitute differences in cell-surface structures that affect each individual and that may therefore be suspected of participating in differentiation. In the case of surface antigens that distinguish one cell type from another, there are....reasons to enquire into function" (ibidem).

There is no *a priori* limitation of the type of function which differentiation antigens could perform. These could be recruited among surface enzymes, hormone receptors, surface molecules triggering some differentiation and/or morphogenetic events (as may be the case for the H-Y antigen), or surface components involved in cell-cell recognition, etc.... It is thus remarkable that most of the differentiation antigens whose function has been

296

revealed by the technique reviewed here seems to belong to the
class of "cell adhesion molecules" (CAM), as it has been defined
for neural tissues of the chick embryo (see Rutishauser *et al*,
1978c) (H-Y of course would make an outstanding exception).
Various features of the immunological approach to the study of
differentiation may account for this result.

(a)   Most of the immunological reagents used are xenogeneic
antisera (the anti H-Y serum again being an exception to the rule).
There is of course no reason why the surface components most
important for the differentiation of a chick embryonic cell should
also be the most potent immunogens when this cell is introduced
into a rabbit. The fact that very few polymorphic differentiation
antigens, against which allogenic antisera can be raised, are
known at the present time to be involved in some differentiation
sequence (as an example, see the experiments by Dulbecco *et al*
with the Thy-1-1 antigen) is of dubious significance. It could
mean simply that the blocking of these antigens by the appropriate
antibodies has not been tried systematically. Alternately, it is
conceivable that polymorphism is more likely to affect structures
of minor functional importance. Also relevant in this context is
the observation made by Dulbecco *et al* that conventional (poly-
clonal) antisera interfered poorly with the differentiation process they were studying when compared with monoclonal antisera.

(b)   Most of the antisera used were raised against whole
cells or cell membranes. The diversity of antigenic determinants
which can contribute to elicit the production of antibodies is
thus quite large. Although nothing is known of the relative
immunogenic potential of the various differentiation antigens
born by the immunizing cells in any of the systems studied here,
it seems clear that the antibodies which react with adhesion
molecules should have an effect epistatic to those antibodies
which recognize differentiation antigens, functioning among
closely apposed cells only. The existence of such differentiation
antigens is little more than a hypothesis at the present time.
It may be important to recognize that the technique evaluated here
will be to a large extent of little help in identifying them
unless immunological reagents devoid of anti-CAM antibodies be
used.

In most of the instances reported here, the effect of antibodies is a broad, morphological one: cells are dissociated, or
a characteristic morphological organization does not take place.
Molecular events concomitant with this organization are seldom
studied however, because of the lack of suitable markers. The
feasibility of such a study would be an important step towards
the understanding of the processes at stake. In particular,
this might help in many cases answering the often obscure question
of which comes first: morphogenesis or genetic determination
(Bennett *et al*, 1971).

ACKNOWLEDGMENT

We thank Drs Rolf Kemler, Dominique Morello, Jean-Francois
Nicolas and Francois Hyafil for permission to mention some of
their unpublished results. The work done at the Institut Pasteur
was supported by grants from the Centre National de la Recherche
Scientifique, the Délégation Générale à la Recherche Scientifique
et Technique (No. 77.7.0966), the Institut National de la Santé
et de la Recherche Médicale (No. 76.4.311.AU), the Fondation pour
la Recherche Médicale Francaise and the Fondation André Meyer.

REFERENCES

*BABINET, C., KEMLER, R., DUBOIS, P. & JACOB, F. (1977) Les fragments
monovalents d'immunoglobulines de Lapin anti-F9 empêchent la formation de
blastocystes chez la souris. C.R. Hebd. Seance. Acad. Sci. Paris, 284,
1919-1922.*

*BENNETT, D., BOYSE, E.A., & OLD, L.J. (1971) Cell surface immunogenetics
in the study of morphogenesis. Lepetit Colloquium, Holland Publisher.*

*BENNETT, D., MATHIESON, B.J., SCHEID, M., YANAGISAWA, K., BOYSE, E.A.,
WACHTEL, S.S. & CATTANACH, B.M. (1977) Serological evidence for H.Y.
antigen in Sxr, XX sex-reversed phenotypic males. Nature, 265, 255-257.*

*BERNARD, O., RIPOCHE, M.A. & BENNETT, D. (1977) Distribution of maternal
immunoglobulins in the mouse uterus and embryo in the days after
implantation. J. Exp. Med., 145, 58-75.*

*BEUG, H., GERISCH, G., KEMPFF, S., RIEDEL, V. & CREMER, G. (1970) Specific
inhibition of cell-contact formation in Dictyostelium by univalent antibodies. Expl. Cell Res., 63, 147-158.*

*BEUG, H., KATZ, F.E. & GERISCH, G. (1973a) Dynamics of antigenic membrane
sites relating to cell aggregation in D. discoideum. J. Cell. Biol. 56,
647-658.*

*BEUG, H., KATZ, F.E., STEIN, A. & GERISCH, G. (1973b) Quantitation of
membrane sites in aggregating Dictyostelium cells by use of tritiated
univalent antibody. Proc. Natl. Acad. Sci. USA, 70, 3150-3154.*

BEUTLER, B., NAGAI, Y., OHNO, S., KLEIN, G. & SHAPIRO, I.M. (1978) The HLA-dependent expression of testis organizing H-Y antigen by human male cells. Cell, 13, 509-513.

BONNER, J.T. (1947) Evidence for the formation of cell aggregates by chemotaxis in the development of the slime mold D. discoideum. J. Exp. Zool., 106, 1-26.

BOYSE, E.A. (1979) Problems of development: immunology as master and servant. Current Topics in Develop. Biol., 13, XV-XIX, Academic Press, New York.

BOYSE, E.A. & OLD, L.J. (1969) Some aspects of normal and abnormal cell surface genetics. Ann. Rev. Genet., 3, 269-290.

BOZZARO, S. & GERISH, G. (1978) Contact sites in aggregating cells of Polysphondylium Pallidum. J. Mol. Biol., 120, 265-279.

BRACKENBURY, R., THIERY, J.P., RUTISHAUSER, U. & EDELMAN, G.M. (1977) Adhesion among neural cells of the chick embryo. I. An immunological assay for molecules involved in cell-cell binding. J. Biol. Chem., 252, 6835-6840.

BROWN, D.D. & GURDON, J.B. (1977) High-fidelity transcription of 5S DNA injected into Xenopus oocytes. Proc. Natl. Acad. Sci. USA, 74, 2064-2068.

CHANG, I.I., REITHERMAN, R.W., ROSEN, S.D. & BARONDES, S.H. (1975) Cell surface location of discoidin, a developmentally regulated carbohydrate binding protein from Dictyostelium discoideum. Exp. Cell Res., 95, 135-142.

COHEN, E.P. & LIANG, W. (1979) Antibody effects on membrane antigen expression. In "Immunological approaches to embryonic development and differentiation I". Current Topics in Devel. Biol., 13, 281-303.

CRITCHLEY, D.R. (1979) Glycolipids as membrane receptors important in growth regulation. In "Surfaces of normal and malignant cells", R.O. Hynes, ed., Wiley & Sons, New York.

CROISILLE, Y. (1970) Applications of immunochemical techniques to the study of certain problems arising from tissue interactions during organogenesis. In "Tissue interactions during organogenesis". E. Wolff, ed., Gordon & Breach, N.Y.

CULLEN, S.E., BERNOCO, D., CARBONARA, A.O., JACOT-GILLERMOD, H., TRICHIERI, G. & CEPPELLINI, R. (1973) Fate of HLA antigens and antibodies at the lymphocyte surface. Transplant. Proc., 4, 1835-1847.

DAVIS, J.C. (1978) Morphogenesis by dissociated immature rat testicular cells in primary culture. J. Embryol. Exp. Morph., 44, 297-302.

DUCIBELLA, T. (1977) Surface changes of the developing trophoblast cell. In "Development in Mammals", Vol.1 (M.H. Johnson, ed.) Elsevier North Holland, Biomedical Press, Amsterdam, New York and London.

DULBECCO, R., BOLOGNA, M. & UNGER, M. (1979a) Differentiation of a rat mammary cell line in vitro. Proc. Natl. Acad. Sci. USA, 76, 1256-1260.

DULBECCO, R., BOLOGNA, M. & UNGER, M. (1979b) Role of Thy-1 antigen in the in vitro differentiation of a rat mammary cell line. Proc. Natl. Acad. Sci. USA, 76, 1848-1852.

DUNIA, I., NICOLAS, J.F., JAKOB, H., BENEDETTI, E.C. & JACOB, F. (1979) Junctional modulation in mouse embryonal carcinoma cells by Fab fragments of rabbit anti-embryonal carcinoma cell serum. Proc. Natl. Acad. Sci., USA, 76, 3387-3391.

EBERT, J.D. (1950) An analysis of the effect of anti-organ sera on the development. J. Exp. Zool., 115, 351-362.

EBERT, J.D., TOLMAN, R.A. & ALBRIGHT, J.F. (1955) The molecular basis of the first heartbeats. Ann. N.Y. Acad. Sci., 60, 958-985.

EICHWALD, E.J. & SILMSER, C.R. (1955) Untitled communication. Transp. Bull. 2, 148-149.

FELLOUS, M., GUNTHER, E., KEMLER, R., WIELS, J., BERGER, R., GUENET, J.L., JAKOB, H. & JACOB, F. (1978) Association of the H-Y male antigen with β2-microglobulin on human lymphoid and differentiated mouse teratocarcinoma cell lines. J. Exp. Med., 148, 58-70.

FRAZIER, W.A., ROSEN, S.D., REITHERMAN, R.W. & BARONDES, S.H. (1975) Purification and comparison of two developmentally regulated lectins from D. discoideum. J. Biol. Chem., 250, 7714-7721.

GELTOSKY, J.E., SIU, C.H. & LERNER, R.A. (1976) Glycoproteins of the plasma membranes of D. discoideum during development. Cell, 8, 391-396.

GELTOSKY, J.E., WESEMAN, J., BAKKE, A. & LERNER, R.A. (1979) Identification of a cell surface glycoprotein involved in cell aggregation in D. discoideum. Cell, 18, 391-398.

GERISCH, G. (1961) Zellfunctionen und Zellfunctionwechsel in der Entwicklung von D. Discoideum. V. Stadienspecifizische Zellkontaktbildung und Ihre quantitative Erfassung. Expl. Cell. Res., 25, 535-554.

GIUDICE, G. (1962) Restitution of whole larvae from disaggregated cells of sea urchin embryos. Dev. Biol., 12, 233-247.

GIUDICE, G. & MUTOLO, V. (1970) Reaggregation of dissociated cells of sea urchin embryos. Advances Morphol., 8, 115-158.

GOLDBERG, E.H., BOYSE, E.A., BENNETT, D., SCHEID, M. & CORSWELL, E.A. (1971) Serological demonstration of H-Y (male) antigen on mouse sperm. Nature, 232, 478-480.

GRAHAM, C. (1977) Teratocarcinoma cells and normal mouse embryogenesis. In "Concepts in Mammalian Embryogenesis". (M.T. Sherman, ed.), M.I.T. Press, Cambridge, Mass., 315-394.

GUSTAFSON, F. & WOLPERT, L. (1967) Cellular movement and contact in sea urchin morphogenesis. Biol. Rev. Cambridge Philos. Soc., 42, 442-498.

HAKOMORI, S. (1975) Structures and organization of cell surface glycolipids. Dependency on cell growth and malignant transformation. Biochim. Biophys. Acta. 417, 55-89.

HANDYSIDE, A.H. (1978) Time of commitment of inside cells isolated from preimplantation mouse embryos. J. Embryol. Exp. Morph., 45, 37-53.

HORSTADIUS, S. (1973) Experimental embryology of echinoderms. Clarendon Press, Oxford.

HUESGEN, A. & GERISH, G. (1975) Solubilized contact sites A from cell membranes of Dictyostelium Discoideum. FEBS Letters, 56, 46-49.

HYAFIL, F., MORELLO, D., BABINET, C. & JACOB, F. (1980) A cell surface glycoprotein involved in the compaction of embryonal carcinoma cells and cleavage-stage embryo. In preparation.

HYNES, R.O. (1976) Cell surface proteins and malignant transformation. Biochim. Biophys. Acta. 458, 73-107.

JACOB, F. (1978) Mouse teratocarcinoma and mouse embryo. Proc. R. Soc. London. B., 201, 249-270.

JOHNSON, M.H., CHAKRABORTY, J., HANDYSIDE, A.H., WILLISON, K. & STERN, P. (1979) The effect of prolonged decompaction on the development of pre-implantation mouse embryo. J. Embryol. Exp. Morph., 54, 241-261.

KEMLER, R., BABINET, C., EISEN, H. & JACOB, F. (1977) Surface antigen in early differentiation. Proc. Natl. Acad. Sci. USA, 74, 4449-4452.

KESKI-OJA, J., MOSHER, D.F. & VAHERI, A. (1976) Dimeric character of fibronectin, a major fibroblast surface associated glycoprotein. Biochem. Biophys. Res. Comm., 74, 699-706.

KOHLER, G. & MILSTEIN, C. (1975) Continuous cultures of fused cells secreting antibody of predefined specificity. Nature, 256, 495-497.

LEUNG, C.C.K. (1977) Embryotoxic effects of heterologous antisera against rat Reichert's membrane. J. Exp. Zool., 200, 295-302.

LINGWOOD, C.A. & HAKOMORI, S. (1977) Selective inhibition of cell growth and associated changes in glycolipid metabolism induced by monovalent anti-bodies to glycolipids. Exp. Cell. Res., 108, 385-391.

LINGWOOD, C.A., NG, A. & HAKOMORI, S. (1978) Monovalent antibodies directed to transformation sensitive membrane components inhibit the process of viral transformation. Proc. Natl. Acad. Sci., 75, 6049-6053.

LOOMIS, W.F. (1975) Dictyostelium discoideum: A developmental system. Academic Press Inc., New York.

MA, G.C.L. & FIRTEL, R.A. (1978) Regulation of the synthesis of two carbo-hydrate-binding proteins in Dictyostelium discoideum. J. Biol. Chem., 253, 3924-3932.

MABUCHI, I. & OKUNO, M. (1977) The effect of myosin antibody on the division of starfish blastomeres. J. Cell Biol., 74, 251-263.

MAUTNER, V.& HYNES, R.O. (1977) Surface distribution of LETS protein in relation to the cytoskeleton of normal and transformed cells. Cell Biol., 75, 743-768.

McCLAY, D.R. (1979) Surface antigens involved in interactions of embryonic sea urchin cells. Current Topics Develop. Biol., 13, 199-214.

McCLAY, D.R. & HAUSMAN, R.E. (1975) Specificity of cell adhesion: differences between normal and hybrid sea urchin cells. Dev. Biol., 47, 454-460.

McCLAY, D.R., CHAMBERS, A.F. & WARREN, R.H. (1977) Specificity of cell-cell interaction in sea urchin embryo. Appearance of new cell surface deter-minants at gastrulation. Dev. Biol., 56, 343-355.

McCLAY, D.R. & CHAMBERS, A.F. (1978) Identification of four classes of cell surface antigens appearing at gastrulation in sea urchin embryo. Dev. Biol., 63, 179-186.

McCLAY, D.R. & MARCHASE, R.B. (1979) Separation of ectoderm and endoderm from sea urchin Pluteus Larvae and demonstration of germ layer specific antigens. Dev. Biol., 71, 289-296.

MEIZEL, S. (1978) The mammalian sperm acrosome reaction, a biochemical approach. In "Development in Mammals", Vol.3, Martin Johnson, ed. pp.1-65, North Holland Publishing Company, Amsterdam.

MINTZ, B. (1965) Experimental genetic mosaicism in the mouse. In: "Preimplantation stages of Pregnancy", pp.194-207. G.W. Wolstenholme & M. O'Connor, eds., J. & A. Churchill, London.

MONOD, J. & JACOB, F. (1961) Teleonomic mechanisms in cellular metabolism, growth and differentiation. Cold Spring Harbor Symposium, XXVI, 389-401.

MOSCONA, A.A. (1957) The development in vitro of chimeric aggregates of dissociated embryonic chick and mouse cells. Proc. Natl. Acad. Sci. USA, 43, 184-189.

MOSCONA, A.A. (1961) Rotation-mediated histogenetic aggregation of dissociated cells: a quantifiable approach to cell interactions in vitro. Exp. Cell Res., 22, 455-475.

MULLER, K. & GERISCH, G. (1978) A specific glycoprotein as the target of adhesion blocking Fab in aggregating Dictyostelium cells. Nature, 274, 445-449.

MULLER, W.E.G., MULLER, I., PONDELJAK, V., KURELEC, B. & ZAHN, R.H. (1978a) Species-specific aggregation factor in sponges. Isolation, purification and characterization of the aggregation factor from Suberites domuncula. Differentiation, 10, 45-53.

MULLER, W.E.G., ZAHN, R.K., KURELEC, B. & MULLER, I. (1978b) Species-specific aggregation in sponges. Transfer of a species-specific aggregation receptor from Suberites domuncula to cells from Geodia cydonium. Differentiation 10, 55-60.

MULLER, U., ASCHMONEIT, I., ZENZES, M.T. & WOLF, U. (1978a) Binding studies of H.Y. antigen in rat tissues. Indication for a gonad-specific receptor. Hum. Genet., 43, 151-157.

MULLER, U., ZENZES, M.T., BAUKNECHT, T., WOLF, U., SIEBERS, J.W. & ENGEL, W. (1978b) Appearance of hCG-receptor after conversion of newborn ovarian cells into testicular structures by H-Y antigen in vitro. Human Genet., 45, 203-207.

MULLER, U., SIEBERS, J.W., ZENZES, M.T. & WOLF, U. (1978c) The testis as a secretory organ for H-Y antigen. Human. Genet., 45, 209-213.

NACE, G.W. (1955) Development in the presence of antibodies. Ann. N.Y. Acad. Sci., 60, 1038-1055.

NAGAI, Y., CICCARESE, S. & OHNO, S. (1979) The identification of human H-Y antigen and testicular transformation induced by its interaction with the receptor site of bovine fetal ovarian cells. Differentiation, 13, 155-164.

NEW, D.A.T. & BRENT, R.L. (1972) Effect of yolk-sac antibody on rat embryos grown in culture. J. Embryol. Exp. Morph., 27, 543-553.

NEWELL, P.L. (1977) Aggregation and cell surface receptors in cellular slim moulds. In "Receptors and Recognition series B, Vol.3", J.L. Reissig, ed. Chapman & Hall, London.

NICOLAS, J.F., KEMLER, R. & JACOB, F. (1980) Effects of anti-embryonal carcinoma serum on aggregation and metabolic cooperation between terato-carcinoma cells. In preparation.

NOLL, H., MATRANGA, V., CASCINO, D. & VITORELLI, L. (1979) Reconstitution of membrane and embryonic development in dissociated blastula cells of the sea urchin by reinsertion of aggregation-promoting membrane proteins extracted with butanol. Proc. Natl. Acad. Sci. USA, 76, 288-292.

OHNO, S., NAGAI, Y. & CICCARESE, S. (1978) Testicular cells lysostripped

302

of H-Y antigen organize ovarian follicle-like aggregates. Cytogenet. Cell Genet. 20, 351-364.

OHNO, S., NAGAI, Y., CICCARESE, S. & SMITH, R. (1979a) In vitro studies of gonadal organogenesis in the presence and absence of H-Y antigen. In Vitro, 15, 11-18.

OHNO, S., NAGAI, Y., CICCARESE, S. & IWATA, H. (1979b) Testis-organizing H-Y antigen and the primary sex-determining mechanism of mammals. Rec. Prog. Horm. Res., (in press).

PARISH, R.W., SCHMIDLIN, S. & PARISH, C.R. (1978) Detection of developmentally controlled plasma membrane antigens of D. discoideum cells in SDS-polyacrylamide gels. FEBS Letters, 96, 366-370.

POOL, R.R. (1979) The role of algal antigenic determinant in the recognition of potential algal symbionts by cells of Chlorohydra. J. Cell. Sci., 35, 367-379.

RAY, J., SHINNICK, T. & LERNER, R.A. (1979) A mutation altering the function of a carbohydrate binding protein blocks cell-cell cohesion in developing Dictyostelium discoideum. Nature, 279, 215-221.

REITHERMANN, R.W., ROSEN, S.D., FRAZIER, W.A. & BARONDES, S.H. (1975) Cell surface species-specific high affinity receptors for discoidin: developmental regulation in D. discoideum. Proc. Natl. Acad. Sci. USA, 72, 3541-3545.

RICHARDSON, C.L., BAKER, S.R., MORRE, D.J. & KEENAM, T.W. (1975) Glycosphingolipid synthesis and tumorigenesis. A role for the Golgi apparatus in the origin of specific receptor molecules of the mammalian cell surface. Biochim. Biophys. Acta, 417, 175-186.

ROSEN, S.D., HAYWOOD, P.L. & BARONDES, S.H. (1976) Inhibition of intercellular adhesion in a cellular slime mould by univalent antibody against a cell surface lectin. Nature, 263, 425-427.

ROSEN, S.D., KAFKA, J.A., SIMPSON, D.L. & BARONDES, S.H. (1973) Developmentally regulated carbohydrate-binding protein in D. discoideum. Proc. Natl. Acad. Sci. USA, 70, 2554-2557.

ROSEN, S.D., CHANG, C.M. & BARONDES, S.H. (1977) Intercellular adhesion in the cellular slime mould: Polyspondilium pallidum inhibited by the interaction of asialofetuin or specific univalent antibody with endogeneous cell surface lectin. Dev. Biol., 61, 202-213.

RUNGGER, D. & TURLER, H. (1978) DNAs of simian virus 40 and polyoma direct the synthesis of viral tumor antigens and capsid proteins in Xenopus oocytes. Proc. Natl. Acad. Sci. USA, 75, 6073-6077.

RUNGGER, D., RUNGGER-BRANDLE, E., CHAPONNIER, C. & GABBIANI, G. (1979) Intranuclear injection of anti-actin antibodies into Xenopus oocytes blocks chromosome condensation. Nature, 282, 320-321.

RUTISHAUSER, U., THIERY, J.P., BRACKENBURY, R., SELA, B.A. & EDELMAN, G.M. (1976) Mechanisms of adhesion among cells from neural tissues of the chick embryo. Proc. Natl. Acad. Sci. USA, 73, 577-581.

RUTISHAUSER, U. & EDELMAN, G.M. (1978) Transmembrane control and cell surface recognition. In "Differentiation and Development", Miami Winter Symposia, 15, 211-233. Academic Press.

RUTISHAUSER, U., THIERY, J.P., BRACKENBURY, R. & EDELMAN, G.M. (1978a) Adhesion among neural cells of the chick embryo. III. Relationships of the surface molecule CAM to cell adhesion and the development of histotypic patterns. J. Cell. Biol., 79, 371-381.

RUTISHAUSER, U., EINAR GALL, W. & EDELMAN, G. (1978b) Adhesion among neural cells of the chick embryo. IV. Role of the cell surface molecule CAM in the formation of neurite bundles in cultures of spinal ganglia. J. Cell Biol., 79, 382-393.

RUTISHAUSER, U., BRACKENBURY, R., THIERY, J.P. & EDELMAN, G.M. (1978c) Cell adhesion molecules from neural tissues of the chick embryo. In "The Molecular Basis of Cell-Cell Interaction". R. Lerner & D. Bergsma, eds., Birth Defects Original Article series, Vol.XIV, 305-316, Liss., New York.

SCHEER, U., SOMMERVILLE, J. & BUSTIN, M. (1979) Injected histone antibodies interfere with transcription of lampbrush chromosome loops in oocytes of Pleurodeles. J. Cell. Sci. 40, 1-20.

SELA, B.A. & EDELMAN, G.M. (1977) Isolation by cell-column chromatography of immunoglobulins specific for cell surface carbohydrates. J. Exp. Med., 145, 443-449.

SIEBERS, J.W., PETERS, F., ZENZES, M.T., SCHMIDTKE, J. & ENGEL, W. (1977) Binding of human chorionic gonadotrophin to rat ovary during development. J. Endocrinol., 73, 491-496.

SIGURDSSON, B. (1940) Antigenic properties of tissue living cells. Proc. Soc. Exptl. Biol. Med., 45, 237-242.

SOLTER, D. & KNOWLES, B. (1975) Immunosurgery of mouse blastocyst. Proc. Natl. Acad. Sci. USA, 72, 5099-5102.

SPIEGEL, M. (1955) The reaggregation of dissociated sponge cells. Ann. N.Y. Acad. Sci., 60, 1056-1078.

SPIEGEL, M. & SPIEGEL, E.S. (1975) The reaggregation of dissociated embryonic sea urchin cells. Amer. Zool., 15, 585-606.

STEINEMANN, C., HINTERMANN, R. & PARISH, R.W. (1979) Identification of a developmentally regulated plasma membrane glycoprotein involved in adhesion of Polysphondylium pallidum cells. FEBS Letters, 108, 379-384.

SZULMAN, A.E. (1975) Experimental in vitro organogenesis and its modification by antibody directed to a cell surface antigen. Develop. Biol., 43, 101-108.

TAKEISHI, M. (1977) Functional correlation between cell adhesive properties and some cell surface proteins. J. Cell. Biol., 75, 464-474.

TAKEISHI, M., OZAKI, H.S., TOKUNAGA, K. & OKADA, T.S. (1979) Experimental manipulation of cell surface to affect cellular recognition mechanisms. Develop. Biol., 70, 195-205.

TARKOWSKY, A.K. & WROBLESKA, J. (1967) Development of blastomeres of mouse eggs isolated at the 4 and 8 cell stage. J. Embryol. Exp. Morph., 18, 155-180.

THEILER, K. (1972) The House Mouse. Berlin: Springer Verlag.

THIERY, J.P., BRACKENBURY, R., RUTISHAUER, U. & EDELMAN, G.M. (1977) Purification of a cell-adhesion molecule from neural retina. J. Biol. Chem. 252, 6841-6845.

URUSHIHARA, H., OZAKI, H.S. & TAKEICHI, M. (1979) Immunological detection of cell surface components related with aggregation of chinese hamster and chick embryonic cells. Develop. Biol. 70, 206-216.

WACHTEL, S.S. (1979) Immunogenetic aspects of abnormal sexual differentiation. Cell, 16, 691-695.

WACHTEL, S.S., KOO, G.C., OHNO, S., GROPP, A., DEV, V.G., TANTRAVAHI, R., MILLER, D.A. & MILLER, O.J. (1976) H-Y antigen and the origin of XY female wood lemmings (Myopus schisticolor). Nature, 264, 638-639.

WACHTEL, S.S. & KOO, G.G. (1980) H-Y antigen. In "Mechanisms of Sex Differentiation". Academic Press, London.

WILEY, L.M. & CALARCO, P.G. (1975) The effects of anti-embryo sera and their localization on the cell surface during mouse preimplantation development. Develop. Biol., 47, 407-418.

WILLISON, K.R. & STERN, P.L. (1978) Expression of a Forssman antigenic specificity in the preimplantation embryo. Cell. 14, 785-793.

YAMADA, K.M. (1978) Immunological characterization of a major transformation-sensitive fibroblast cell surface glycoprotein. Localization, redistribution, and role in cell shape. J. Cell. Biol., 78, 520-541.

YAMADA, K.M. & PASTAN, I. (1976) Cell surface protein and neoplastic transformation. Trend. Biochem. Sci., 1, 222-224.

YAMADA, K.M., YAMADA, S.S. & PASTAN, I. (1976) Cell surface protein partially restores morphology, adhesiveness and contact inhibition of movement to transformed fibroblasts. Proc. Natl. Acad. Sci., USA, 73, 1217-1221.

YAMAIZUMI, M., UCHIDA, T., OKADA, Y. & FURUSAWA, M. (1978a) Neutralization of diphteria toxin in living cells by microinjection of antifragment A contained within resealed erythrocyte ghosts. Cell, 13, 227-232.

YAMAIZUMI, M., FURUSAWA, M., UCHIDA, T., NISHIMURA, T. & OKADA, Y. (1978b) Characterization of the ghost fusion method: a method for introducing exogenous substances into cultured cells. Cell Struct. Funct., 3, 293-304.

YAMAIZUMI, M., UCHIDA, T., MEKADA, E. & OKADA, Y. (1979) Antibodies introduced into living cells by red cell ghosts are functionally stable in the cytoplasm of the cells. Cell, 18, 1009-1014.

YANAGIMACHI, R., WINKELHAKE, J. & NICOLSON, G.L. (1976) Immunological block to mammalian fertilization: survival and organ distribution of immunoglobulin which inhibits fertilization in vitro. Proc. Natl. Acad. Sci. USA, 73, 2405-2408.

ZENZES, M.T., WOLF, U. & ENGEL, W. (1978a) Organization in vitro of ovarian cells into testicular structures. Hum. Genet., 44, 333-338.

ZENZES, M.T., WOLF, U., GUNTHER, E. & ENGEL, W. (1978b) Studies on the function of H-Y antigen: dissociation and reorganisation experiments on rat gonadal tissue. Cytogenet. Cell. Genet., 20, 365-372.

ACKNOWLEDGEMENTS

We acknowledge with gratitude permission to reproduce figures as follows:

Academic Press, Developmental Biology, Vol.74, C.W. Lo & N.B. Gilula (1980), I, Figures 1,2,5a,5b and Table 2; II, Figures 3a and 5c.

M.I.T. Press, Cell, Vol.18, 399-409, C.W. Lo & N.B. Gilula (1979), Figures 3,4 and 6.

M.I.T. Press, Cell, Vol.18, 411-422, C.W. Lo & N.B. Gilula (1979), Figures 1,2,4,6,7 and 11.

Academic Press, Developmental Biology, Vol.69, 247-257, Wartiovaara et al. (1979), Figures 1A,1B,3A,3B,4B,5B,5C,6A and 6B.

Academic Press, In: The Cell Surface: Mediator of Developmental Process (Ed. S. Subtelney), Wartiovaara et al. Figures 11, 12,14a and 14b.

Spring-Verlag, Differentiation 10 (1978), I. Thesleff Figure 13.

Academic Press, Cell Interactions in Development (Eds. M. Karkinen-Jääskelainen, L. Saxen & L. Weiss). Chapter by I. Thesleff, Figure 7.

Company of Biologists, J. Emb. Exp. Morph. 34 (1975), E. Lehtonen, Figure 8.

Rockefeller University Press, J. Cell Biol. 76, 748-760, 1978 (Figure 12, p.756) and 81, 83-91, 1979 (Figure 6, p.88 and Figure 7B, p.89).

SUBJECT INDEX

chondrogenesis in somitic mesenchyme,
see somitic mesenchyme

chondroitin 4 - and 6-sulphates, in
intercellular matrix of notochord and spinal cord, 172

chordamesoderm: duffusible substances from, in induction
of neuroectoderm, 164-7, 170

chorioallantoic membrane, muscle
development in chick limb
buds grafted to, 144

choriocarcinoma, 92

chorion, basement membrane in, 242
fibronectin in, 243

chymase, acts on fibronectin,
241, 254

cleft palate: induction of, in
rodent foetus by mid-
gestational doses of
glucocorticoids to mother,
208-9
H$_2$ locus and susceptibility
to, 213-15
mechanism of glucocorticoid
effect, 216-20
number of glucocorticoid
receptors, and suscepti-
bility to, 210

colcemid (depolymerizes microtubules), 28
effect of, on cumulus-oocyte
coupling, 29

colchicine (depolymerizes microtubules), 28

collagen
action of collagenase on, 82
in basement membranes, 81, 179-80
in corneal stroma, 174, 175
in development of secondary
palate, 207
in division of muscle masses, 150
in extracellular matrix of notochord and spinal cord, 172
interaction of fibronectin and,
236, 238, 240, 253
precipitated by heparin-
fibronectin complexes, 255
stimulates morphogenesis, 151
substrate for plasmin only after
collagenase action, 84
types I-III (interstitial): cells
secreting, also secrete
fibronectin, 252
type II: fibronectin disappears
from presumptive limb cartilage
at time of appearance of, 258
type IV: distribution of, in early
mouse embryo, 256

collagen synthesis
ascorbic acid and ferrous iron
required for, 252
glucocorticoids and, 207, 219
inhibited by proline analogue, 219
inhibition of, in cell cultures,
does not affect fibronectin
synthesis, 252
simulated by presence of collagen,
172, 175, 176, 179

collagenases
products of action of, on collagen,
84

310

cathepsin G, acts on fibronectin, 241, 254

cell adhesion
fibronectin in, 257, 260
substances involved in,
indentified by antibodies,
296; in check embryonal
neural cells, 274-6; in
Dictyostelium, 279-82; in
preimplantation mouse
embryo, 270-3; in sea
urchin embryos, 276-9

cell death
in development of secondary
palate, 203, 204, 218; prevented in palate cultures by
excess of epidermal growth
factor, 220
lacking in myogenesis in muscular dystrophy? 152
in morphogenesis of tissues, 148

cell interactions, morphogenetic,
161-2
by diffusion of signal substances,
162-4; in lens induction,
167-9; mechanisms of, 169-70;
in neural induction, 164-7
through cell contact, 163, 164,
183; in kidney tubule
induction, 183-8; mechanisms
of, 188-94
through intercellular matrix, 163,
164, 170-1; in corneal stroma
formation, 174-5; in odontoblast differentiation, 175-9;
roles of various matrix substances in, 179-83; in somite
chondrogenesis, 171-4

cell junctions, different types
of, 40
see also desosomes, gap junctions,
tight junctions

cell migration
fibronectin and, 240, 259, 260
implies breakdown of fibronectin-
containing matrices, 254

cells
exchange of substances between different types of, 58-61; *see also*
cell interactions
shape and size of, and number of gap
junctions between, 57-61

cerebrospinal fluid, fibronectin
in, 234

chick embryo
cell adhesion substances from
neural cells of, identified by
antibodies, 274-6
changes in fibronectin distribution
in, during development, 258
development of limb in, 102-4; *see
also* polarizing region
dorsal and ventral muscle masses in
development of limbs of, 143
heart development in, inhibited by
anti-heart antibodies, 269
ionic coupling of cells in, 54
muscular dystrophy in, 152
polydactylous mutants, 129-30

chick-quail chimeras, in experiments
on limb development, 119, 139-40

choline: passage of, from cumulus cells
into oocyte, 12-13, 15, 16, 17;
hormonal effects on, 19

in early mouse embryo: by cytoplasmic bridges, 48, 50, 51;
by gap junctions (from 8-cell
stage), 58
in other early embryos, 54-6
in post-implantation mouse
embryo, 66, 67-8, 69

iron, ferrous: required for synthesis of collagen, 252

kidney tubule induction
contact of epithelial ureter bud
and surrounding mesenchyme in,
183-8
fibronectin disappears from
mesenchyme during, and appears
in tubule basement membrane,
258
inhibited by polyanions, 190
and kidney morphogenesis, 192-4

laminin, glycoprotein specific for
basement membrane
antisera against fibronectin may
contain activity against, 250
in developing kidney tubules,
192, 193
distribution of, in early mouse
embryo, 256

lens
diffusible substances in
induction of, 167-9
influences corneal differentiation by contact, 174

LETS, see fibronectin

limb
apical endodermal ridge in development of, see apical endodermal
ridge
axes of, 102
development of, in chick and mouse,
102-4
distal deepening in, 128
dorsal and ventral muscle masses in,
143-4
effect of placing impermeable
barrier across bud of, 128
fibronectin disappears from presumptive cartilage of, at time of
appearance of type II collagen,
258
malformations of, caused by glucocorticoids, 210
polarizing region in development of,
see polarizing region
progress zone in, 119, 120;
polarizing zone remains at edge
of, 126-7
without muscles, developed from
somatopleural mesoderm grafted
to coelomic cavity, 142, 149

lobster, ionic coupling of cells in
early embryo of, 54

luteinizing hormone (LH)
and cumulus-oocyte coupling, 16, 21,
23
induces granulosa cells in ovary in
vitro (by stimulation of prostaglandin synthesis?), 90
meiosis in oocyte after release of,
13-14
See also gonadotrophins, pituitary

318

lysosomes, form in cumulus cells
and oocytes after loss of
coupling, 22

macroglobulin, alpha-2 (protease
inhibitor in plasma): and
plasmin, 84

macrophages, appear in limb bud
in absence of polarizing
region, 121

malignant cells, fibronectin
and, 241-2, 291-2

mammals
polarizing region in limb buds
of, 106
possibility of culturing whole
embryos of, 155

matrix (extracellular, intercellular, interstitial)
in cell cultures: electron
micrographs of, 237, 238;
not formed by malignant cells,
241; of teratocarcinoma cells,
248, 250, 251; two types of
fibril in, 253-4
composition of, 253
in early mouse embryo, 242, 243,
246
fibronectin in adhesion of cell
surfaces to, 257
fibronectin-containing, formation and breakdown of,
252-5
fibronectin types in, 236, 240
glucocorticoids, and synthesis
of components of, 207

in morphogenetic cell interactions,
163, 164, 170-1, 207; in corneal
stroma synthesis, 174-5; in
odontoblast differentiation,
275-9; in somite chondrogenesis,
171-4; role of various substandes
in, 179-83

mesenchyme
cells of, in developing limb: gap
junctions between, 125; respond
to polarizing region, 119
contact with, in epithelial differentiation, 39-40
dental, odontoblast differentiation
from, 177
presumptive in chick embryo, disappearance of fibronectin from,
258
presumptive limb-forming: antero-
posterior axis of, fixed before
limbs appear, 126
responsible for mutation to polydactyly in chick, 129
see also somitic mesoderm

mesoderm of mouse embryo, 242
fibronectin in cells of, and in
matrix surrounding, 243; between
endoderm and, 245
matrix glycoproteins in, 256

mesonephros
cells of, have low polarizing
activity for developing limb,
118, 122, 126
of chick embryo: specific parts of,
react with anti-Forssman antibodies, 289-90